Friedrich Fröbel, Josephine Jarvis

Friedrich Froebel's Education by Development

The Second Part of the Pedagogics of the Kindergarten

Friedrich Fröbel, Josephine Jarvis

Friedrich Froebel's Education by Development
The Second Part of the Pedagogics of the Kindergarten

ISBN/EAN: 9783337130695

Printed in Europe, USA, Canada, Australia, Japan

Cover: Foto ©berggeist007 / pixelio.de

More available books at **www.hansebooks.com**

INTERNATIONAL EDUCATION SERIES

FRIEDRICH FROEBEL'S

EDUCATION BY DEVELOPMENT

THE SECOND PART OF THE PEDAGOGICS OF THE KINDERGARTEN

TRANSLATED BY
JOSEPHINE JARVIS

NEW YORK
D. APPLETON AND COMPANY
1899

COPYRIGHT, 1899,
BY D. APPLETON AND COMPANY.

ELECTROTYPED AND PRINTED
AT THE APPLETON PRESS, U. S. A.

EDITOR'S PREFACE.

IN the former volume * nearly one half of the essays brought together by Wichard Lange in a volume entitled The Pedagogics of the Kindergarten have already been printed, in Miss Jarvis's translation. Those essays relate more especially to the plays and games, although in several articles the gifts are discussed with some degree of thoroughness. In the present volume the educational principles underlying the gifts are more thoroughly discussed. Again and again in the various essays Froebel goes over his theory of the meaning of the ball, the sphere, the cube, and its various subdivisions. The student of Froebel has great advantage, therefore, in reading this volume, inasmuch as Froebel has cast new light on his thought in each separate exposition that he has made. Sometimes the briefest mention may prove the most illuminat-

---

* No. XXX of the International Educational Series.

ing, and certainly every brief summary helps to understand the extended treatise.

Froebel proceeds from the solid to the surface through tablets and stick-laying, and finally reaches drawing. He returns to the solid through paper folding and the constructing of outlines of the regular solids by means of sticks joined by means of soaked peas.

The essays on the training school for kindergartners and the method of introducing children's gardens into the kindergarten are very suggestive and useful. In fact, there is no other kindergarten literature that is quite equal in value to the contents of this present volume. The remaining essays in Lange's volume not yet translated are mostly of an ephemeral character, treating of occasions like the play festival at Altenstein (No. 29), a speech at the opening of the first kindergarten in Hamburg (No. 28), a sketch of the constitution of a proposed educational society (No. 25), and two other papers of like character (Nos. 23 and 24).

With the publication of the present volume a complete list of the original works of Froebel in English translation has been provided in this series, namely:

Froebel's Education of Man, Vol. V.

The Mottoes and Commentaries of Mother-Play, Vol. XXXI.

The Songs and Music of the Mother-Play, Vol. XXXII.

The Pedagogics of the Kindergarten, Vols. XXX and XLIV.

Besides these, the series furnishes other helpful volumes for the understanding of Froebel, namely:

Miss Blow's Symbolic Education, Vol. XXVI, and Letters to a Mother, Vol. XLV.

Mr. Hughes's Froebel's Educational Laws, Vol. XLI.

W. T. Harris.

Washington, D. C., *January 14, 1899.*

CONTENTS.

CHAPTER	PAGE
I.—THE SPIRIT OF THAT TRAINING OF THE HUMAN BEING WHICH EDUCATES BY DEVELOPING DEMONSTRATED BY THE WAY IN WHICH LINA LEARNED TO READ	1
II.—THE CHILD'S LOVE OF DRAWING	55
III.—GUIDE TO PAPER-FOLDING	89
IV.—STICK-LAYING	118
V.—FROEBEL'S FUNDAMENTAL PRINCIPLES OF EDUCATION, HIS MEANS AND MODES OF EDUCATION AS WELL AS EDUCATIONAL AIM AND OBJECT, IN RELATION TO THE TENDENCIES AND REQUIREMENTS OF THE TIME—REPRESENTED BY HIMSELF	161
VI.—THE FATHER'S CRADLE SONG	215
VII.—THE CHILDREN'S GARDENS IN THE KINDERGARTNERS	217
VIII.—TRAINING SCHOOL FOR KINDERGARTNERS	228
IX.—ADDRESS BY FROEBEL BEFORE THE QUEEN OF SAXONY.	241
X.—THE CONNECTING SCHOOL.	268
XI.—COMPENDIOUS DESCRIPTION OF THE KINDERGARTEN GIFTS AND OCCUPATIONS	306

ANALYSIS OF THE CONTENTS BY THE TRANSLATOR.

I. Spirit of Education by Development: (p. 1) Froebel requested to make an understandable written statement of his system, and its means, methods, objects, and aim; his attempt to do so not quite satisfactory; his desire to find cause of failure; (2) verbal statements understood, recognized as true, and partially carried out; cause of misapprehension; *verbal* statements accompanied by an object; foundation of his educational whole; (3) object connected with explanation; How Lina Learns to Read; (4) significance of child's impulse to learn to write and read; child feels itself a part-whole; (5) good results of education depend on this feeling; (6) it is the starting-point of the kindergarten; how awakened; writing and reading a connecting bond; (7) connects the single with the general; letter-writing; (8) answers to questions about life and education; (9) total result; Pestalozzi and Froebel compared; all-sided life-union; (10) aim and goal; evils of overleaping intermediate steps; (11) means of avoiding this error; (12) Lina's mother our teacher; direct natural attraction; (13) means provided suited to child's powers; (14) teaching founded on Lina's wish; (15) developed from her feeling of personality; (16) problem of developing education solved; (17) error in education; foundation of developing education; (19) child's personality a means of perception; (20) series of development repeated; (21) nature of Creator made known by creation; (22) development implies something to develop; what? (23) education must be faithful to laws of development; what are they? Lina's mother answers; (24) consequence of child's feeling itself a part-whole; what has a developing influence also a part-whole; what is to be developed in child? (25) what are the laws of development? how God revealed his nature; (26) the creation to manifest the divine; impulse and attraction; child's impulse to represent his own nature; education a part-whole; consequence of this; (27) effect of misconceptions on education; Nature a touchstone; Nature and the free spirit; means of criticism; (28) why laws of Nature and the spirit explain one another; the touchstone for developing education; (29) touchstone of correct following of training; (30) mother complies with Lina's wish to learn to write; first step; second step; (31) what is speaking? the word? third step; secret of developing education; law of connection; (32) training continued; child led to find out different sounds; (33) man connects Creator and creation; language, man, and things; elements of speech also connections; law of connection essential; (34) and practical; (35) nature of education by development; what and how to develop; (36) means of testing laws of education; law of opposites; (37) what can be done by means of law of connection? four answers; fifth, education suited to the time; (38) sixth, with what education is connected; seventh, to what it corresponds; it unifies; (39) eighth, law of the triune life; ninth, how developing education is suited to the time; (40–41) personality the point of germination of developing education; (42) starting-point for all education; this point resembles a seed; it constitutes the nature of developing education; (43) unity and universality opposites; (44) effect of closely uniting them; union of opposites made objective; effect of name; (45–46) aim of developing education; of the kindergarten; germinating point recognized; (47) the other developing laws follow; fact in back-ground of mother's management; law of the original unit; lies in man; divineness of his nature; (48) means of testing named; education a science; an art; a living fact; (49) how to obtain all this by developing education; practical understanding; education a finished whole; how to apply in developing education the laws before explained; (50) the particular conditioned by the general; child in combination with the great life-whole; each of its actions not isolated; (51) three purposes of education; meaning of all done by and with child; first attention to him linked with development of his limbs, etc.; (52) he strives for free use of his members; his imitation promoted by the mother; activity of limbs, etc.; (53) effect of word heightened by rhythm and tone; Mother-Play and Nursery Songs; (54–55) a test of this book.

II. Man a Creative Being: (p. 56) man in interdependence with Nature; what this view gives to education; (57) how God reveals himself; how man makes known his being; man determined to create; (58) a young child's spontaneous expressions of life; child a creative being; therefore related to his Creator; (59) reason for his activity; he invests with life all he sees; (60) more proofs that he is related to God; his desire for representing; how he shows his creative power; (61) first object of his tendency to activity; character of his other first playthings; use of these a proof of his creative impulse; his efforts to draw important; in second or third year, bulky solids replaced by other things; (62) these things named; child's creative representations advance as his means of play become less material; (63) illustrations; what man recognizes in and by the spirit; effect of fostering child's creative power; (64) child's desire for typical representation; object of first part of this chapter; foundation of child's activity up to his seventh year; (65) upon what his action depends; his slight power no obstruction to his impulse toward creative activity; his effort to strengthen this impulse not to be disturbed; (66) he seeks material with which to gratify it; outward fostering of this impulse indispensable; parents to give their children what Nature gives hers; (67) why painting and drawing attract the child; his choice of material shows him in harmony with Nature's doings; (68) child's efforts to prove he is a creative being are sacred; he thus shows himself also a member of the great whole; what he is destined to do; (69) why he proves himself a creating being, especially by his drawing; development of child's power of drawing essential to developing education; (70) effect of omitting drawing; development of child produced by his drawing; (71) what it requires from him; its importance; (72) what it makes possible; symmetrical development required; effect of a free position of child's body; fostering of feeling of pleasantness; (73) also of that of the right essential to cultivation of child as a creative being; character of the drawing; (74) sequence of lines drawn; drawing of lines connected with the movements of the limbs; (75) reasons for connecting word and deed; illustrated by the drawing; the way lines originate to be considered; (76) effect of adding the loving tone to the word; lines, material for representation; illustrations; (77) child tries to draw a house; demands upon him made by such drawing; (78) the eye a measure; drawing in the network; (79) the shortest line the measure; longest line five times as long as shortest; how to make the network; (80) distance of parallel lines from one another; conscious drawing begins with the straight line; sequence of such lines; more directions; effect of the network; (81) way to make child conscious of what he has done; observant action the expression of child's activity; (82) drawing conditioned in his nature; to what this first drawing leads; laws of formation; next requirement; (83) effect of employing the rule experienced by child; laws of formation now shown, higher than those before stated; each form suggests the next; capacities aroused in child by his desire for signs; how he feels the growth of his creative power; (84) to what use he turns the law of development; inventiveness a proof of creative power; what child learns by looking at objects; the progress he has made at this stage; (85) effect on him of drawing simple lines; (86) how to develop his power of creating by drawing; the most important result of drawing; child's conscious advance from the straight to the curved line; round and straight compared; (87) what child represents by the drawing; why he is set in the midst of the life-whole; (88) first condition for his development; his cultivation for creative drawing; point of reference of true education; kindergarten leads to this; nature of kindergarten.

III. Paper Folding: (p. 89) what it is; beneficent property of paper folding; (90) A, guide to means of employment in general; means of development conditioned in child's nature; (91) development and employment connected; third to sixth gifts; tablets; what they form; (92) for what they are adapted; sticks; their use; (peas work); interlacing; jointed slat; (93) intertwining; weaving; dividing and recombining material; form changed, quantity unaltered; modeling; folding; (94) thread game; cutting; (95) what this review shows; (96) from what paper folding proceeds; square preferred to triangle; B, paper folding a means of employment; how the square may be made; what the results of this work should show; squares formed from rectangles; (98) reader to *do* the work; (98–99) directions; (100) squares material for folding; development of forms; children make their own squares; aim of folding; geometric surfaces named according to the number of their sides, etc.; (101) directions for fundamental forms; law at the foundation of all education; (102) directions; child to do and hear; (103) why; perceptions of form and size gained by the folding; examples; (104) this is play to the child; double meaning of phrase "right angle"; facts cleared by repetition; (105) from what the eighth [geometric] perception proceeds; fact stated in the opposite way; (106) more perceptions; (107) more directions; (108) perceptions; (109) twofold meaning of "right angle"; (110) the right angle a measure of angles; a perception; child perceives many facts; eye and mind developed; (111) directions; double truth derived from results of folding; what *doing* the thing proves; a perception twice shown; (112) directions; (113) perceptions; concluding statement; (114) opposite square; perceptions; kindergartner gives the word for child's perception; (115) concluding directions; (116–117) final perceptions.

IV. Stick Laying: (p. 118) the whole of plays and occupations not arbitrarily originated; by what called; how grown; what it is; what comes from it; (119) developing education the fundamental effort at present; stage of child's development required for stick laying; (120) result of bringing it forward too soon; from what the sticks proceed; (121) tablets split into sticks; effect; later develops from earlier; development of sticks in play-whole; (122) from what the sticks, etc., result; three principal directions; (123) how they appear in ball, etc.; sticks a means of cultivation; ball compared to the germ of a tree; (124) to a flower bud; what this is to show; importance of early leading the child into the linkings of life; (125–126) stage of cultivation required for use of sticks; (127–128) what is gained by this presentation of the stick play; (129) connecting laws united in the stick; it has the essential properties of all objects; (130) its relations to other objects; its possible positions; cause of its charm for the child; (131) what it is for him; author and reader enter a kindergarten; number of children there; their greeting; that of kindergartner; (132) Froebel's answer; her apology for the simplicity of the intended stick-play; the manifold goes forth from the simple; illustration; (133) questions about shape and position of stick asked and answered; it is compared with other objects; (134) she gives each child a stick; tells what her own is like, and lays it on table; each child does the same; she repeats the names given to the sticks; (136) children show where each imagined object is; what the children do and say proves that manifoldness proceeds from the single; the children's joyousness; (137) their mutual sympathy; effect; farewell song; what Froebel learned in the kindergarten; (138) permission to come again; reader's desire to return shows interest in the subject; (139) he has thought of it constantly since visiting the kindergarten; Froebel's experience similar; (140) sticks compared to magnet; (141) attraction of each part of the play-whole explained; (142) effect of inner power abiding in Nature; (143) innermost nature of relation between kindergartner and children; author and reader enter the kindergarten; sticks again; attention song; each child has one stick; (144) is given another; effect of visitors joining in the play; influence of the invisible; (145) no activity without effect in the kindergarten; result of kindergarten fostering; children name their stick forms; reasons for laying the forms on the table; (146) kindergartner makes angles by beckoning; (147) three kinds of angles; their names; children make angles by beckoning; kindergartner lays angles with sticks; children do the same; (148) angle song; other stick forms; visitors asked to test kindergartner's knowledge; (149) they do so; she names kinds of angles in some of the stick forms; elasticity play with sticks; result of play rhymed; (150) play repeated; vertical and horizontal illustrated by steel yard made with sticks; the use of former gifts gave knowledge of angles; why ways are given for gaining this knowledge by stick-laying; (151) everything in a double relation to the child; reason why kindergarten material satisfies the child; (152) recognition of inner phenomena shows the true nature of kindergarten; simple material best liked; deeper insight into relations between child and material required; (153) advice to mothers, etc.; (154) reader derived benefit from the kindergarten; inner connection of forms made by the children; (155) what these forms show; that which is invisible yet perceptible; (156) visitors look at children's stick forms; children name forms; different names for the same form; (157) why allow children to make forms not much like the real objects; kindergartner counts forms; children look closely at them; sticks taken up; each child remakes what forms it can remember; (158) effect of this; aim of kindergarten; (159) connecting power in children's minds differs; objects made are named by kindergartner in a stick story; effect of such story telling; (160) harmony originated in youth, re-presented in later life; farewell to the children; taking up the sticks, a reward; farewell song.

V. Principles of Education: (p. 161) man drawn to observe the phenomena and facts of his own time; attention to its *character;* a view of the whole; education characterizes the time; (162) cause of this; periodicity recurs with humanity viewed as a whole; a period of development begins now; neglect to give it special attention; (163) this neglect explained by analogy; neglect of development has painful results; how the understanding of life is cleared; in what it expresses itself; man's whole life one of education; by what does the present time prove itself a time of education? (164) facts on which the answer to this question rests; (165) universality and consciousness of endeavors required; both present in this period of time; individual requirements of this period; (166) first demand which characterizes the time as educational; to what it appertains; (167) to what pressure toward self-comprehension has given rise; man's effort to raise to consciousness his tendency to activity; man thinking, feeling, and acting; second demand; (168) the mother to be first raised to recognition of her dignity, etc.; third and fourth demands; (169) fifth and sixth demands; relation of especial to general most important; to what it leads; (170) the State an educational institution; seventh and eighth, by what and why the present time is characterized as educational; (171) thought and idea afford what life and man require; how onward development of the individual as a part-whole is obtained; keystone of the whole— ninth; (172) in what man's vocation lies; effect; these strivings the most essential of the new period of life; *all* are present *at the same time;* (173) *all* must be fulfilled at the same time; this problem to be solved in the same way as a gardener educates his plants; from what this manner of treating the child proceeds; (174) two facts derived from observation of life; these facts give two of the maxims on which true education is founded; third maxim; (175) what these two limitations give as a product; fourth law of education; those who acknowledge the truth of these principles must begin to carry them out; woman's instinct to be raised to consciousness; (176) women the first trainers of the human being; man has no less part in this training; this co-operation to begin in childhood; why; humanity composed of opposites; child to be treated in accordance with the spirit of this system; (177) where the source of genuine development is; when and how it finds nourishment; how nurses quiet a young child; how child lulls itself; (178) rhythm and song connected with child's expressions of life; so they belong to a healthy education of man; effect of song on child; Mother Play and Nursery Songs mostly accompanied by song; effect; singing tone added to word; (179) with such development, child needs an object by which to develop himself; human being's nature requires a counterpart; effect of not providing one; (180) whither the images of fancy lead; what kind of object to provide for child's activity; what he chooses; what girls like; (181) why; weight the first expression of attraction; what the object must be; (182) why ball is best liked; (183) play engendered by the opposite like; (184) how ball can be used; in what form and relations; (185) ball an all-sided means of development; child seeks for plurality; human gardener to bring to unfolding child's longing for plurality; why; child seeks in plurality the connecting unity; this is given by the colors of the balls; (186) how different number and like form, size, etc., of ball increase with child's increasing age; (187) fact that ball is child's true first plaything disputed; proofs of fact; (188) ball a means of moral preservation; (189) what child's activity proves; sphere the opposite like of ball, and so child's next plaything; it is compared with ball; (190) progress in material in accordance with Nature; meaning of last phrase; the germ of the later must lie in the earlier; (191) requirements of education explain one another; this proves the truth of the whole; sphere not to supplant ball; each gift extends the use of the preceding one; (192) use of sphere has much in common with ball play; how sphere may be moved by string; each plaything a whole; each has its task to do in child's education; task of sphere; why we need clear perceptions of unity in life; (193) effect of leading the child to this; proof of the divine in life; how sphere is to benefit child; effect of rhymed song in child's education; (194) effect of using white and black spheres; next plaything must be a contrast to the sphere; (195) cube the third plaything; its contrasts to the sphere; law of connection approached with ball; (196) it now shows itself to be a law of life; for what it is later essential; how it is to be brought to child's notice; another respect in which the perception of such a law is important; (197) what we, as true educators, must give again; why; return to cube; to what it leads by its form; how it may be introduced; (198) form, etc., important; also the sure gaze; why and how the child should be early introduced into the perception of form, etc.; cube leads to relations of number; (199) third way of viewing the cube; why it is especially suited for play; man born for research; to what cube leads and introduces the child; (200) with what his delight in the object of play is connected; what takes place within him is developed during the play; kind of plaything which the child likes best; (201) what gives value to the representation plays; what they are; these representations a whole, though incomplete; how to complete it; value of modes of play, etc., here demonstrated; (202) reasons for this value; the peace which Jesus left must become a fact; what he said must be true; (203) these remarks applied to child life; what has been said about the cube, the keystone to the kindergarten system; law of connection the most important law of the universe; child to be treated in accordance with the highest laws of life; child *is* life; his plays, etc., represent life; (204) cylinder the connection between cube and sphere, and fifth object of play; child life proves the truth of this last; (205) requirements followed in the choice of these three early playthings give the same result; what experience proves with regard to them; by what the use of the cylinder is determined; sphere, cylinder, and cube, a whole; what they form; they point toward the phenomena of art life; example; (206) tripartite character of the columnar orders; purpose of bringing out this fact here; (207) what must be done if human beings are to unite in truth; spirit of union compared to the sun; (208) how to gain all-sided life union; we must have a clear idea of the nature of our subject before we can safely go on; the object of play was always a whole; three objects considered as wholes, form another whole; (209) they are therefore part-wholes; the property of being a part-whole, important for the child; he can not too early be led to observe it; this is done by next plaything; instinct of mother and that of child lead to this; child desires a separable whole; the next gift is a cube divided into eight equal cubes; (210) knowledge of cube form important for life; illustrations; (211) the attraction these eight cubes have for children proved by experience; (212) a few lines will serve for guidance to those who are urged by deep earnestness to test all that is revealed for child's education; separation requisite for the observing intellect and the outwardly representing life; apparent separation is inner union; why the man can not learn to live in life; (213) the comprehension of the original unit fostered and observed in the child's life in the way of educating children which lies before us; the second and third gifts fulfill this requirement perfectly, yet meet with the most hostility; what proves that they develop child life most judiciously; the most striking proof of the comprehensiveness of this genuine training; (214) answer to the reproach that children are earnest in kindergarten; why children should rejoice over Froebel's plays.

VI. The Father's Cradle Song (p. 215); (216) song concluded.

VII. The Children's Garden in the Kindergarten: (p. 217) acquaintance with Nature the sure foundation of successful education; Nature the direct manifestation in action of God; Nature's growth and development to be compared with man's; (218) the child to have opportunity for this comparison in the children's garden; why the kindergarten requires a garden; child to prove himself by action an indvidual member of a greater life; (219) the children's beds in their garden to be surrounded by the garden of the whole; aim of children's garden; child compares the plants in it; (219-221) suggestions for its general arrangement; (221-226) particular arrangement; (226) retroactive effect and influence on the child of such fostering of plants; (227) even the tending of a window garden benefits the child.

VIII. Plan for Training School: (p. 228) what kindergartens are; 1, general aim of the institution; (229) 2, aim of the institution in particular; (230) 3, forming plays for the designated aim;

(231) 4, age of those who enter; 5, stage of cultivation of those who enter; (232) 6, duration of the training course; 7, the attainment of the aim of cultivation; (233–235) 8, division of time during the training course; (236–238) 9, a few more essential particulars and the keystone of the training course; (238) 10, outside conditions of entrance; 11, beginning of the course; (239–240) 12, concluding remarks; (240) 13, reference.

IX. Address in Dresden : (p. 242) in Nature, all is in that inner coherence which leads to God ; facts of Nature and life demonstrate this ; in Nature, the manifestation of God, all is intuition and life ; this coherence deeply grounded ; (243) what living in this coherence gives man ; since those highest ideas of life are represented in Nature, should not man strive to live in, and to lead the children into, harmony with this coherence? (244) what the earnestness of this question caused Froebel to do ; if there is coherence in Nature and life, each individual must be a whole and a part of a whole ; (245) why child is such a part-whole ; how is he to rise to the anticipation, etc., of this coherence? how he should be treated from the first; he is in coherence with God, Nature, and humanity ; errors in regard to this threefold comprehension of child injure his unfolding; this triune conception of child, our first problem ; second, to what he is to be led ; (246) what the foundation of all development is; child's life expresses itself by activity ; point at which a satisfactory education begins ; activity threefold ; in what it appears ; child's life to be treated according to this triplicity ; Nature leads child to this; apple a part-whole ; (247) Nature too near to and too far from the human being ; a connecting third needed ; ball fulfills the conditions of this third ; to what it corresponds ; how ; (248) ball the representative of all which exists ; foundation of ball's connecting child and Nature ; what man must do to understand Nature ; Froebel's plays a means of introduction into Nature ; ball unites opposites in itself ; it is an introduction into the knowledge of Nature and man ; (249) it is a mirror of all ; what this intimation about ball's nature justifies ; the inflexible proceeds from the movable ; sphere and cube come from ball ; (250) what sphere and cube show ; three principal directions in sphere and ball the key to recognition of every form, etc. ; three activities in each object; importance of this observation of a coherent three ; (251) the ends of the three principal directions are surfaces in cube ; cube's surfaces and corners postulated in its interior ; all directions alike in size in sphere and ball, different in cube ; (252) it makes its inside externally visible ; what the outside of the sphere is ; the great

ANALYSIS OF THE CONTENTS. xxi

law of Nature and life; why it is important to express this law; the highest aim of man's activity; (253) child to be guided to this in his play; ball the means of introduction into the general, cube into the particular; illustration; (254) child-fostering, the representation of knowledge of Nature; why child thankfully recognizes such fostering; why retain for recognition the fountain, etc., of the phenomena of life; aim of consideration of Nature and genuine child-fostering; (255) object of giving sphere and cube to child as the next plaything; each of these subjects important for the fostering of child's life; indivisibility requires the divided; child shows this; (256) requisites for next plaything met by third gift; attention given to the eighth part-cubes; opposites to appear united; the one-sided a development of all-sidedness; how does the third gift correspond to child's nature; (257) tendency of intelligence and heart in early and later life compared; form which can be separated and united is demanded by life and furnished by the divided cube; this makes it possible to comprehend the child as feeling, thinking, and creating; forms which can be made with third gift are either those of knowledge, of beauty, or of life; (258) to what this corresponds; the three principal directions in sphere and cube compared; (259) these directions to become abiding and different; how this is done; effect of this division on formation; forms made with fourth gift are of the same kinds as those of the third; (260) how each series is extended; effect of having the three directions differ in size in the building stones; child impressed by what he can do with these small means; deduction; (261) effect of such impressions; each form admits of several perceptions; why these observations are important for the child; effect of these plays; (262) general law exemplified; this law as important as its opposite; child led by Nature and play to recognize it; what he perceives and to what he is led; (263) why the twice divided cube is the next (fifth) gift; difference of kind in its parts as well as increase in their number required; how this is done; (264) correct comprehension of the right important for the child; the former plays sought to confirm Nature's impressing of the right; oblique and inclined also important; examples; to what these plays train the child; what this gift does; (265) kinds of forms made with it; these forms compared with those of the former gifts; (266) Pythagorean problem illustrated with this gift; special importance of this gift for the child; its opposite (sixth gift); character of the forms made with the latter; why the demonstration of these plays closes here; what it was possible for Froebel to do; (267) the progress and conclusion of the whole; application of these plays; Nature the foundation of human education; what would be possible if Froebel's wish were fulfilled.

X. The Connecting School: (p. 268) Emma wants to know how to manage a connecting school; different stages of child's development; childhood separated into the baby and family stages; (269) how child develops in the latter; kindergarten the second stage; the human training of the kindergarten; here a child comes to a plurality of objects; (270) what they become to and teach the child; third stage, connecting school; kindergarten acquirements; the training stage of kindergarten sharply bounded; abstract knowledge first entered upon in connecting school; (271) name shows nature; it combines kindergarten and school for learning; result of kindergarten; with what connected; child comes to the perception of manifoldness in unity; child introduced into the science of the general and special laws of life; (272) why Emma's perception of the stage of kindergarten training must be decided; law of development counteracts impulse to destruction; result of this developing activity; to what stick laying, etc., leads; surface laying preceded that of sticks; to what the former leads; what it teaches; kindergarten nature, etc., to be reproduced in an organic manner; (274) the being right and the whole kindergarten based on mathematical proofs; two more perceptions of life begin in kindergarten; an important fact in relation to even and uneven numbers; where number first finds its true recognition; (275) what feeling should be strengthened in Emma; what is demonstrated by "How Lina Learns to Read"; effect of clearly producing this subject; (276) what exercises belong in connecting school; what completes kindergarten cultivation; (277) nature of kindergarten, school, and connecting school; (278) what the latter forms; comprehension of its nature, etc., not easy; why; length of connecting school training course; why such schools are so rare; with what such a school connects and what it gives; (279-280) examples; into what connecting school leads; reference to "Education of Man," etc.; the most important means of passing from kindergarten to school; (282) keystone of kindergarten employment; box of fourteen solids; such forms to be made in clay, etc.; use of this box in kindergarten; with what education by development begins; why complete development of limbs, etc., is needed; what the child is to learn to know; why ball serves this purpose; to what it leads; (284) what ball demands; what sphere requires; of what sphere and cube are the expression; cylinder shows the connection between the two;

ANALYSIS OF THE CONTENTS. xxiii

employment with the fourteen solids is connected with the second gift; objects of this gift the four first solids; (285) what these objects have shown the child; inner law of change results from law of connection; where this law came forth; into what contemplation of the fourteen solids introduces the child; (286-287) manner of making the transition from cube to octahedron; (288) transition from cube to dodecahedron; their places in the box; (289) by what the three surface directions of the cube lead to the sphere; effect of suppressing corners; places in box of new solids and their completing forms; these six new solids close the course in kindergarten; reverse course also given there; (291) outlined forms important at the kindergarten stage; children made the outlined square, etc.; from what the outlined cube results; how the next solid originates within the cube; the outlined octahedron; (292) outlined tetrahedron; all these to be represented in the cube; comparison of these with each other and with the solids gives an intimation of what? child ready to enter connecting school; keystone of kindergarten, starting point of connecting school; (293) list of opposites; child skilled in these contrasts; (294) they are to be brought to his notice in sequence; leading direction not lacking; germ for each development provided; connecting school develops child from unconsciousness to consciousness; nature of connecting school; examples; (295) general result; what determines the form of a body; introduction into the science of space, etc.; (296) with what consideration of the fourteen solids is connected; analysis of cube; instruction in number to be carried from the stage of perception to that of conception; (297) also instruction about the form and size of solids, etc.; to what perception of figure leads; consideration of the outer world; to what it leads; (298) province of language; tone and rhythm; to what song leads; science of plants connected with that of the surface of the earth; illustrations; (299) helps in geography; examples; what is connected with this; (300) why observation of plants is important for connecting school child; (300-301) this importance shown by Bible texts; plant world important to Germans; cultivated fruit trees; what they are; (302) human being to be the same; how he can become so; from and into what he can pass with this anticipation; what the connecting-school teacher must have before her eyes; it is doubtful if she can fully apply to her school what has been said; (303-304) Froebel convinced of possibility of connecting school; places in box of the fourteen solids and their completing forms; (305) into what these solids lead; their principal divisions.

XI. Kindergarten Means of Employment: (p. 306) man a sentient being; with what and by whom connected; the child a member of a family; the foundation of his development; what position he obtained by such fostering, etc.; (307) how the mother appears through such fostering; with whom and what she connects her child; how she must act with regard to all these connective offices; why they are of equal importance; (308) child to develop himself at a future time; how and between what he is to mediate; by what means he first develops himself; what mediation presupposes; (309) child and Nature opposite yet alike; child a part of Nature; how he must develop; difference between man and Nature; care for child's health the first duty of mother or nurse; (310) when he is healthy; why he is given a body; limbs and senses are contrasts; so are arms and legs; (311) between what the hands are a connection; to what the senses correspond; physical treatment of the child; what limbs and senses require; what to observe in reference to child's development; (312) development of his impulse to activity; this impulse soon requires an object; the retroactive effect on the child a twofold one; what the first plaything must be; (313) these requirements met by the ball; what essential properties of objects are represented by it; of what it is a means; development produced by ball play; (314) play with ball as a type; go back to earliest childhood; what child's increased power requires; the opposite of the sphere described; sphere and cube opposite yet alike; (315) connection required; cylinder described; cone required and described; why these four form a whole; (315-316) comparison of these solids; (317) mode of playing with them; results of moving them; effect on child; to what the play with them leads; into what it introduces the child; of what these objects are the source; (318) why this gift has been opposed; what Froebel considers it; he offers to prove his assertion; return to development of gifts; playthings hitherto undivided; child likes opposites; (319) why his impulse to create must be fostered; next plaything the divided cube; how and by what required; (320) by what third gift plays are conditioned; two rules for its use; threefold character of forms made with it; what these forms are called; this distinction important; representations connected with word and melody if possible; (321) guidance for the use of this gift; what the three principal directions are in this, and what they must be in the next gift; (322) how the fourth gift results from the third; variety of forms made with fourth gift; forms also of three kinds; (323) fifth gift compared with third in

ANALYSIS OF THE CONTENTS. xxv

regard to the oblique line of direction and the number of divisions; (324) forms made with fifth gift also of three kinds; their effect; sixth gift parallel to fifth; shape and number of blocks in sixth gift; its peculiarity; its likeness to the fifth gift; seventh gift results from fifth; (325) what all its parts put together form; oblique surfaced equal parts; polyhedrons represented as developed from the cube; eighth gift related to seventh; first series of children's plays; first and second series of children's playthings; (326) third and fourth series; conception of surface as independent of the solid appears in the last-named series; four series of tablets; (327) derivation of these plays from the preceding, clear to the thinker; laws prominent in forms of beauty; their starting-point, means of progress, and return; contrasts developed from connection; (328) three kinds of forms made with tablets; exercises in color added; division of tablets gives sticks; these are embodied lines, attract the child, and form a new division of the means of play, etc.; first and second kinds of play with them; (329) third kind; (330) stick plays train the eye; laws of development from within; fourth kind of stick play; (331) fifth kind; recapitulation of kinds of stick plays; points of connection for the singing tone, etc.; (322) why the leader of children must have a clear idea of all this; over what this remark extends; from what points come and what they form; objects used to represent them; point of connection for the collection of natural products as means of play; (333) analysis of solid into surface, line, and point, compared to the development of a tree; hence we must return on our path to the first unity; separation requires coherence; this is obtained by the pricking; materials described; (334) connection of points to lines; the pricking sheet; its peculiarity; three series of pricking sheets; for what and how the last is a preparation; letters connected with the pricking; (335) color connected with pricking; why the results of the pricking should be used as presents; lines combined to form surfaces and solids; interlacing; intertwining; (336) weaving; how the results of it may be used; importance of children's giving; (337) the making of mats and baskets connected with the weaving; peaswork used to make surface forms, outlined solids, and furniture, etc.; (338) the development of forms from the preceding and finally from the original form; why this occupation is important; with what the cutting from wood of sleds, etc., is connected; what can be made by these employments; what results; (339) children's making presents; combination of surfaces; paper surfaces folded into boxes; how to fasten the sides; (340) cardboard modeling; a new division of children's employments; form changed but not quantity; thread lines; stick lines; paper folding; modeling; form changed, quantity diminished; cutting from squares; (341) the form made with straight lines, curved lines, both; free-hand cutting; cutting out; what unfolding takes place here; child uses all his powers; why the busying of children at this stage of development ends here; by what this is shown; (342) change of solids made of soft material; here again is a close; the cutting of solids connected with undivided spheres; divided ones; they are divided in three ways; (343) cylinder divided in four ways; cone also; what proceed from the connection of round and straight; (344) provinces connected with the modeling; what proceeds from all this; collecting of pebbles, leaves, etc.; its effect on the child; to what the collecting of plants, bugs, etc., leads; effect of all this on parents, etc.; (345) children the most enjoyable playmates of a child; why; nature of children's plays; what the child does by means of them; (346) aim of this play-whole; what child discovers in the plays; their effect; foundation of child's impulse to imitation; (347) what are revealed to the child in the mirror of his plays.

EDUCATION BY DEVELOPMENT.

I.

THE SPIRIT OF THAT TRAINING OF THE HUMAN BEING WHICH EDUCATES BY DEVELOPING, DEMONSTRATED BY THE WAY IN WHICH LINA LEARNED TO READ.

I HAVE often been requested to give a written statement of the fundamental truths and principles of my system of fostering and training childhood and youth. I have also been asked to word this statement in such a manner that it can be easily understood. A further request has been made that I should include in this statement an account of the means and methods, object and aim, of this system of training which I call an educational whole, that educates by developing.

I have several times attempted to meet this demand, which seemed to me a just one; but my attempt has never given complete satisfaction to those who desired the statement.

A man generally likes to make himself intelligible concerning that which is the business of his life, and which he has at heart, as soon as it becomes of importance to the public. I have therefore tried to discover the cause of this non-satisfaction, especially as my verbal communications concerning the subject in question have been understood by my hearers.

Indeed, they have told me, times without number, that they fully recognized the truth of my statements, thought about them earnestly, and even partially carried them into practice, though with less clear insight, a smaller degree of perfection, and less understanding of the logical connection.

Aided by such precise statements, I could not fail to find the cause of misapprehension in the different effect which a verbal communication has as compared with a written one. The former is always connected with an object which brings me and my hearers together, and serves as a symbol that helps them to understand my statements. For the thought becomes perceptible, and so at once full of life, by the use of such an object.

This fact shows in a remarkable manner that not only the training and development of children and young people, but that of mankind in general, is (especially in early life) connected not merely with that which is perceptible, but with that which is, at the same time, perceptible and symbolic. The educational whole here presented receives its deep human foundation—a foundation which is also natural and all-embracing—from the important statement just made. That is, the spirit, everlasting but invisible, is made perceptible and recognizable by means of the material through the sense (Sinn) and the symbol (Bild), through the symbolic (Sinnbildliche) as the connection between the spirit and the material.

Now, in order to give an intelligible explanation resting upon this foundation, and which can be generally understood by the people, it is necessary to find out such a general connecting object—such a symbol, as it were. This object is speech as connected with visible signs (symbols made permanent)—it is learning to write and read. For the ability to read and write is now universal (at least in Germany), and therefore what is said about reading and writing must necessarily be universally intelligible to the people—at least I believe and hope so.

Therefore I now choose the presentation in the preceding chapter, " How Lina learns to write and read," * as the connecting and intermediate object of contemplation of my present communication about my educational whole. This selection is the more appropriate as the desire to learn to read and write is a direction of the awakening impulse to-

---

* See chap. xv, Pedagogies of the Kindergarten, vol. xxx, International Education Series.

ward employment and culture, which shows itself quite early in each German child. An educational whole which claims to be comprehensive and to meet the needs of mankind in general must proceed from something which belongs to all humanity. I therefore begin the presentation of the fundamental ideas and principles of this educational whole with this question, What is the signification of the child's peculiar impulse to learn to write and read? and what is the general significance of this impulse in the child and for the child?

It is, in general (according to the stronger or weaker feeling of personality attained by the child), the effort to busy itself in this personality like the observant adults around it; the effort to prove that *it* also is a member of the great general life-whole, and, as far as possible, to introduce itself into and show itself in this life-whole. This same feeling urges the child on to wish to help its father (and still more its mother, who more fully enters into this wish) whenever circumstances allow. This feeling is twofold, for the little one feels itself at the same time a self-poised whole and a part depending on the great whole or totality of life which it perceives around it, and which it divines in itself.

This feeling, or, in other words, this presentiment of itself as a part-whole, certainly does stir in the child, however slightly. I consider the observation, acknowledgment, and fostering of this feeling to be the foundation, the starting point—I might say, the germinating point—the heart and fountain of the true, developing, educating cultivation of the child and of the human being; or, to express it generally and in a single phrase, the education of man in general.

The good results of all true education depend on the careful notice, fostering, development, strengthening, and cultivation of this feeling on the part of the child that he is a whole, and yet also a part of all life; and on the avoidance of every violation, clouding, disturbance of it. It is the point of union of all things, and of each thing which is to be attained through education. Indeed it, singly and alone, first makes possible a true human, all-sided, life-united education. But through it such an education does become possible; for through it the child recognizes itself directly in the two relations of independence and dependence without needing to be instructed by any outward means. Without the direct recognition of those relations there is no genuine human education, as, in nature, no healthy, complete development is possible without observation of those two relations, and without the mute, unconscious living in accordance with these relations. The great, invisible working gardener of the universe, of nature, and of humanity shows us this in the education of all his children, as the active plant gardener does in tending the smallest pot garden. Therefore this twofold feeling or anticipation is the starting point, the vital point of the true kindergarten. It is also the vital point of the developing educational training of the child, of the youth, of the human being, and of humanity up to all-sided life-union. Such training, verified by the kindergarten, is the all-sufficing way leading at once to completeness; or at least preparing for the possibility of, and entering upon the path which leads to completeness.

Through all this the second question obtrudes, How and through what is this feeling awakened on the part of the child of his twofold relation as a part-whole—a feeling which at first slumbers deeply within him, of which so much has been said, and which has been seen to be so important.

This feeling is awakened by almost everything that is done for or with the child. In manifold ways he feels and sees himself (especially through his frequent oppositeness to grown-up people) as a particular and individual thing in contradistinction to the general and collective. But this all-effective feeling is especially awakened when the child is encouraged to self-activity and to a developing busying of himself while with his parents (especially his mother), or at least in the company of real educators. In such company the child soon feels an invisible but uniting bond, which embraces all grown-up people, and even things. He feels a bond which surrounds and unites all things for which he asks. But besides that invisible bond he soon remarks a visible and still more effective bond which connects the farthest as well as the nearest. This is the wonderful art of writing and reading, mute yet speaking, moving men in many ways, and bringing to them joy and sorrow, pleasure and pain, laughter and weeping.

Thus the perception of his twofold relation is aroused in the child by his seeing and perceiving that writing and reading are a means of connecting the separate and single with the general, a means of uniting the single parts with each other. Letter writing especially awakens this perception in the child. What child does not like to write letters? How often the request, "Please, please give me some paper, dear father" (or "dear mother"), "I want to write a letter." George is now just three years old. Some time ago he sent to his father the scribbled or merely folded pieces of paper which he imagined to be letters. When his mother also was obliged to take a journey which would keep her away from him for a long while his most earnest request was, "Mother, write me a little letter." The kind mother agreed to do so, and sent folded sheets like little letters, on which the child's fancy read what suited him and what he expected, as if it were in the letter. This actually took place. Therefore our Lina's desire in the former chapter,

"Please, please, dear mother, give me some paper, I will also write a letter," is in harmony with the great laws of development and education.

Now, what is symbolically expressed in the life picture just brought forward as an answer to the questions before us concerning life and education?

First, the child feels early through his whole being what he is, and, through that which comes to pass around him and exerts an influence as a self-dependent, individual, and separate being, and as an active member of a great life-whole, he divines himself as a part-whole.

Second, he feels in like manner that he can exist only in this life-whole; can develop only through this; can become what he is to be only in life-union with this.

Third, therefore his desire and effort are to show himself as such a part-whole; he wishes and begs to be permitted to occupy himself as such.

Fourth, it is thus the spirit of the surrounding life which acts on the slumbering qualities and capacities (germs, heart centres, and starting points) in the child, as the sun's light, the earth's warmth, the materials of life and nourishment in the air and water act in spring on the seeds, germs, and sprouts of the plants. In the case of Lina in our little story it is the mother's loving fostering, the father's thoughtful notice, and the uncle's requirements and helpful sympathy.

Total result, the child will and must be recognized as a member of the great life-whole. He is to be tended, developed, educated, trained, and treated as such in all-sided union of life. His wishes and expressions, all his indications of life, point toward this view of the child, which is also the meaning of the wish "Teach me to write."

Reference has been made to Pestalozzi's Influence of the Home. He places this influence at the summit of his educational means as the first requisite. My educational means have been brought into comparison with those of Pestalozzi. Comparisons are always favorable to the promotion and application of truth. I will therefore state, in respect to the point mentioned, that, as Pestalozzi claims for his child the influence of the home, so do I claim for mine the powerful might of all-sided life-union which accompanies it from childhood. Thus life-union consists of the management of the child and the observation of the human being in and according to all the relations of life—to the whole life-power—following the pattern of Nature, who treats the smallest seed and the least plant like the entire world-process and realizes God's image in nature.

I return to our Lina and to her learning to read and write in order in this also to perceive (symbolically) what is here stated.

With the feeling of the particular and the general, of the personal independence and of the dependence on the whole, and with the efforts thereby called forth to give itself up to this whole and to this general (as our Lina did by writing, indeed by the writing of a letter), the goal appears before the child at the same time with the wish, and is possessed of attraction and charm. This is proved by the fact that the child overleaps all limitations (means, way, and manner) in order to reach this goal.

That man, in looking toward the aim of his wishes and desires and keeping it in mind, frequently overlooks the proper means, the right path toward, and the best manner of attaining this aim, is a phenomenon and an experience proceeding, indeed, from the child world, but confirmed by all education of individuals as well as of nations and of humanity.

But this hurrying (from the germ to the fruit, from the wish directly to the fulfillment, from the desire to the aim, springing over all the necessary conditions which should be previously fulfilled) has had the saddest and most pernicious results in life in the education of the individual as well as in that of whole communities, in the education of the nations as well as in that of the human race— even in that of all humanity. This haste has had such sad results for the individual that he could not overcome them in the whole course of his life.

Yet we see communities, nations, the human race—yes, all humanity—up to this instant suffering from this single error, which, by its pernicious results, inexorably brings chastisement to man. This is one of the most injurious errors, if not the most injurious one, in the education of the individual as well as of all men.

Yet though it is one of the most dangerous, it is, alas, also one of the least clearly recognized errors in education. But the means to avoid this error (so hurtful in its consequences in the education of children, of individual human beings, as well as of whole nations) are far less recognized and applied. And yet the means are so simple, being the opposite of the error—that is, stability.

What is it that teaches us to know this means in its application, to avoid this error in human education (when recognized) which leads to a constant chronic disease, and is, at the same time, so universally extended? It is the opposite here as always which instructs us by the connection it requires. Thus the opposite of human education, which is the constantly developing education of Nature, teaches us to avoid this error. And how? In Nature the impulse, the arousing and striving, the goal or aim, are always quite near to one another.* The way from the striving to the nearest

---

* Impulse and goal or aim are opposites. The arousing and striving form the connection.—Tr.

goal, from the impulse to the nearest aim, is always very short. For means and aim, way and goal, lie always very close together in the education of Nature; indeed, to the unpracticed eye they often seem to coincide.

Now, who teaches us to employ this means of constant development in human education? Lina's mother, who herself follows thoughtfully as well as consciously the constantly developing education of Nature. She shows us this education in the way in which she teaches writing and (immediately connecting its opposite with it) reading to her little daughter Lina.

First of all, she makes the child notice in a manner which is intelligent and capable of truth that in order to reach a goal or attain an aim conditions must be fulfilled, powers developed, and means appropriated, and that the employment of these means must be practiced beforehand in the right way. All are mere expressions of the mother, repeated and often confirmed to the child by the smallest, first, and best of its own actions and wishes (directly founded on fact). Having this confirmation, the child is not relegated to a remote, uncertain *future* in which the discovery is certain to come to it.

It is quite essential here to notice that with this matter of *direct natural attraction*, which is so important for the child's life, the mother (or the educator in general) meets with aid the endeavors even of the smallest child. For with many things which move before the child (for example, the ball which swings by a thread or string to and fro before it) the child looks (not constantly, but for a very short time) at the appearance of the swinging ball, but it seeks and looks for the cause of this swinging appearance—for the moving hand. This fact supplies the proof (even in the smallest child), arouses and shows (even in the child itself) the anticipation, that <u>man</u> is a being who questions and investigates the causes and origin of appearances and things. This fact is yet further confirmed by several phenomena and facts of the simple instinctive acts of children, about which I will say more by and by when we notice further these spontaneous actions.

We will now go further in our observation of the thoughtful way in which Lina's mother proceeded. We see that her fostering motherly feeling also goes further. She not only places means and object, way and aim, etc., as nearly as possible by one another, but she also makes the means so easily handled, so suited to the powers of the child, to its developing use of limbs and senses, that the employment of these means gives the child but little trouble (easily overcome by inclination and pleasure), since will and deed can thus coincide directly in one action.

All is just as easy for the child to remember as to do, because, according to the already often-mentioned first law of cultivation in nature, here also in the art of writing, as in all art, the completest opposites are to be combined with one another —the living, sounding word and the dead, mute stick; the will of the child and the manageableness of the stick which is without will; the spirit and the material.

The little girl, anticipating all this, although as yet dimly, makes joyously her request, "Mother, teach it to me!"

But from this request (after appropriating the means) the little girl's gaze—confirming our remark made above—springs immediately to the aim, the object. "If any one could only read what I write," says the child sadly, having in view only the aim and object of the writing. "An experiment will test it, the doing will show it," answered the mother simply; thus instructing, educating the child by word and object, deed and explanation, by neither alone, but, as before remarked, by the opposites which are most intimately connected in the whole contemplation and perception, where the immaterial and material again present themselves in the recognition of what has been done in the reading as in all art.

With what does the mother again connect her instruction by word and deed? Or, rather, from what does the mother derive her instruction? Or, yet more precisely, how and from what does she develop the instruction desired by the child herself? First of all, she founds such instruction on the wish of the little girl, and lets it, as it were, grow out of that wish.

Even this taking up and noticing the child's wish by the mother is very important for answering our question. Therein lies the imitation of the above-asserted constancy of Nature (the constancy of the education of Nature in the field, the province of human education and for education). The mother now resembles here in her action the sun, which in spring awakens the slumbering power in seeds and buds, which slowly rousing further nourishes and strengthens itself. And so it is to be with all human education.

Further, the mother develops this instruction, so much desired by Lina, not only from that which is personally experienced by the child, but also from and by means of her own direct feeling of personality. The mother connects her instructions with that which directly arouses this feeling in each human being. She demands the name of one familiar and beloved, the name of the father, lastly the little girl's own name, and so particularly connects the instruction with the child herself. She develops from the child's innermost nature, thus from the point, from the fountain where desire and fulfillment coincide, from the power of the soul, in which will and action are one. Therefore the will and deed of the little girl are in harmony with the wish and the requirement of the mother. For here Lina's mother has solved for us clearly and consciously, and thus completely (as each mother does instinctively with more or less obscurity, greater or less imperfection), the most important problem, the as yet little recognized secret of true developing education and genuine instruction, and she has at once practically applied this solution. She has drawn out both education and instruction in all-sided life-union from the life, the impulse, the wish and the will, the power and the individual activity of the child, as well as from the little girl's self-reliance and self-determination, and has done this by means of the child's own action. The mother's influence thus resembles that of the spring sun, which by warmth awakens the life (the impulse, the power, the self-activity, and the self-determination) in each seed kernel, arouses in it the impulse to unfold according to its natural capacities that which lies in it by its own activity and all-sided union with Nature. It is enough to say that Lina's mother has solved with a word the problem and revealed the mystery. She has also put the solution immediately into practice. She has transformed the education and instruction, which before were foreign to the child, into self-education and self-instruction for her daughter Lina.

Lina's mother here solves with clearness and precision the problem and reveals the mystery of genuine education and instruction, since she turns them into self-education and self-instruction for her child. She has previously shown the harmfulness of springing from the wish to the aim without paying any attention to the intermediate links and requirements, and by observing the constancy in the education of Nature, has disclosed the means for avoiding that harmfulness in human education. She here shows us still further another of the greatest and most injurious failures and wants of our methods of education and instruction up to the present time. Education and instruction, discipline and school, seek, as a rule, the grounds for determining their requirements and their management either wholly outside of the life of the children or, even if within the life of the human being, yet derived from a time which is, in respect to the child (the little charge, the pupil, the scholar), so remote, so far in the future, as to have for him no power at all of attraction, of arousing, and of development. That which the child, the pupil, is to do and learn must proceed from its power of will and action inwardly united to a doing, to a desire, by means of the direct, instantaneous effect of the total life united in itself.

Certainly this is shown by almost all our subjects of instruction, especially as applied to the mass of people. Our instructions in reading and writing, as also in counting and speaking, arithmetic and language, are especially feeble, as they mostly begin with the abstract with which instruction should close; hence the few abiding results of this instruction in life.

Therefore what is to have true, abiding and blessing, instructive and formative effect on the child as pupil and scholar, and as a future active man—viz., independent employment—must not only be founded on life as it actually appears, must not only be connected with life, but must also form itself in harmony with the requirements of life, of the surroundings, and of the time, and with what they offer. It must especially have an arousing and wakening effect on the inner life of the child, and must thus spontaneously germinate from that life. This is the nature of the developing educational training of man, to follow and practice which I regard as the indispensable demand of the time (founded on the laws of Nature and the world, on the necessary laws of all the formations of life), and the maintenance of which I recognize as the demand of life. I hold it in its general comprehensive application as so highly important to the life of humanity and of the nations, that its realization and accomplishment (in proportion to the degree in which it is connected with simple, unchangeable laws) should be the task of all educators, in all relations of life, and under all circumstances. These methods of education and training the kindergartens also represent consciously, and true kindergartners have this object firmly in view, carry it lovingly in their hearts, and strive for it in all that they do.

Let us now further observe the course of training which Lina's mother followed and which was founded on what has been before stated. Proceeding with womanly tact from that which is manifoldly double-sided, the name, the personality, which unites in itself person and thing, creating and receiving (writing and reading), self-employment and learning, etc., becomes again to the child in its manifold double-sidedness a type of herself while she thinks and speaks it, speaks and hears it, hears and writes it; having written, sees and reads it, makes it again audible, and so again leads back to the thought in the mind. We see here the spiritual and corporeal united in one body, and using it as a symbol we comprehend and recognize all which exists and lives. Here the progress has been from existence and life as such, from the spirit, which then as a sequence leads back again to the recognition of existence and life as such, and of the spirit of all which has appeared and is appearing.

And now we see with what deep womanly thought Lina's mother found her course of training not only in a name, but in the name of the child itself, and discovered how the way of training entered upon by herself and the child's own course of development coincide; how the laws, limitations, and requirements of the one are at the same time those of the other, and therefore the requirements of the mother are easy for the child to comprehend and to fulfill.

We further see how the mother has found in the name a means of perception, a symbol, as it were, by which to make the child recognize that the spirit is perceptible in the corporeal, and as it is perceptible in the corporeal can come forth from it and become active.

Through this course of proceeding the mother has also (which is of like importance with the preceding) obtained a means of bringing the child to conceive, perceive, and understand the working of the spiritual upon the spirit, and of even making it perceptible in itself (going through the corporeal, the body, as it were, from the body, acting again upon the spirit, and being felt and recognized by the spirit as spiritual). This takes place when the series of developments before given (that of the awakened thought visibly appearing in written, formed words; from this again come forth to perceptible spiritually reawakening thought) is several times carried through with the child audibly and visibly (comparing inner cause and outer appearance, outer appearance and inner effect), as was the case with Lina in her repeated letter writing.

But now what is the further natural necessary result in simple continuous development of this course of training pursued by Lina's mother? The works speak, the things speak, the nature makes itself known from the form; by the form the spirit manifests itself. By that which has been produced and created the nature and spirit of the producer and creator make themselves known. As the world, the universe appears to be always becoming and constantly creating, it also appears to have become, to be created, it appears as a creation. It must therefore necessarily express, reveal, and manifest the nature of its original cause—the spirit of its Creator. If we now listen to the one great accord of the world, resolving in itself all distances, it sounds "good!" But good, in its completeness and perfection as it appears in the universe, as the harmony of the world, includes in itself the beautiful, the true, and the right. Therefore goodness in itself, and, as it were, complete in itself, must necessarily be the nature of the Creator of the world and of the universe. Nature therefore makes known the being of God; it renders his nature clear and perceptible to us. Thus Nature, being in itself single and also living by its own power, shows and testifies that, in all its rest, it is living; with all its changing existence, it is existing; with all its manifoldness, it is, in itself, single, and is thus the complete expression of the goodness of its Creator.

How highly important for life are these sayings! What deeply grasping life-comprehending truths! How they begin with the simplest and close with the highest! They therefore correspond to the anticipations of the child's mind, as well as satisfy the investigating nature of the man's spirit. They can be developed for the thoughtful child from the nearest, even from its name (as we are taught by our examination of Lina's learning to read and write), and at the same time also show the germinating points of deeper knowledge to the thinking adult.

Indeed, for ourselves the starting point for showing the nature and spirit of the developing, educational training of man and the demonstration of its possible realization comes out by this consideration from symbolic perception. For if there is to be development there must be something to develop; if there is to be education there must be something to draw out, to educate; if there is to be cultivation there must be something to cultivate. The question, therefore, is first, What is there in the child to develop, to educate, to cultivate? That which develops and cultivates itself only according to limitation—that is, according to laws. But education is intertwined with development and cultivation; therefore it must also correspond to the laws of development and cultivation of the one who shapes it. Those who educate must therefore inevitably not only know, but act in conformity with and be faithful to these laws of development of the one who is to be formed by education. A further question, therefore, is, What are the laws of development and cultivation, and what is the test of their being rightly followed?

The course of cultivation followed by Lina's mother gave and gives us in a symbolic view, by its result, the answer to all this and the solution of all this. It showed and taught us that all created things bear within themselves the nature of the Creator; but the Creator is in himself good; the child is also a creation, and therefore also bears within him the nature of his Creator—goodness. And further it taught us how the Original Cause of all life is one and single, bearing life in himself and creating life from himself. So also the life of all which is manifold and apparently isolated in the universe is, according to its inner nature, single. And each individual being as it is in appearance a whole in itself is also, in accordance with its nature, a part of the uniform life of creation, therefore at the same time a part and a whole —a part-whole which, even in its slightest detail, as a separate being, not only feels itself as a part-whole and lives as such, but also shows in its separate existence the life of the whole world and of all Nature. We recognize even in the beginning in Lina's request that the child early felt herself as a part-whole, occupied herself as such, and made claims as such. From this comes the great—that is, all-comprising, and therefore highly important—sequence: As each individual and separate being is a part-whole of the all-life, so also its laws of development are those of the whole world and the whole of life; only they will be manifested in peculiar separate ways and limits, determined by the separateness of the separate being.

But all which is individual, which exerts a developing and thus educating influence upon any separate being in Nature and the creation, in the whole world and in all life, is likewise a whole and a part of the whole of the world and of all life. Consequently, as it bears within it the nature and life of this whole, it develops and acts according to the laws of this whole; but only in its special peculiar way, therefore also according to the laws of development of each individual, deducting that which is determined by its separateness and peculiarity.

And after this the above-stated questions can be answered as simply as precisely:

1. What is that which is to be developed, to be

educated, to be cultivated in the child? It is the nature of the Creator, who is also the Creator of the child; it is the divine nature as it appears limited by creation, world, and Nature, by humanity and the human race, by much separateness, especially by the separateness of personality and individuality.

2. What are the laws of development and formation according to which man is to develop by education? They are the laws of development and cultivation which have their cause and their source in God as the Creator of the world, by which and according to which the world was created, by which and according to which Nature, being and life, goodness and love, reveal themselves and still make themselves known in humanity, in the human race, and in the individual human being, and which therefore appear in each newborn child, living and working anew as essence.

But in order to speak like the child in a childlike, and, as a German and a man, in a German, manlike way, since He who is in himself single and good (in which words, as I have before said, are comprised all the other qualities which are recognized as divine, such as love, life, etc.) revealed his nature from inner self-determination, outwardly demonstrated, and declared, and disclosed his nature as single and good, living and loving. That is, as the Creator of the world created and creates from himself, created the universe, the world, and so the creation, the universe appeared, as it were, to be drawn from within him. Therefore the creation, Nature, and even the human being should from inner self-determination make known and manifest the divine—that is, the nature, being, and life of the Creator of each. And so a feeling single in itself, indeed, but dual in appearance (as each individual, being dual, is opposite to the single), the feeling of impulse and attraction grew in each separate existence, and so above all in man, as a sign and testimony, as it were, that he has his source and origin in the single, is born from the single as individual, and is to feel himself at the same time a part and a whole. Therefore, according to this, the child's slumbering impulse to develop and represent his nature by his own choice and his own determination requires from without the educating attraction to waken and arouse this impulse. This fact is the foundation of the natural, original, reciprocal relation of pupil and educator, which is here only intimated. But since the developing educator is, as well as the pupil, a manifold part-whole of the all-life, etc., he, as such, carries within him (though in a manner peculiar to himself) the general laws of the whole, and especially the laws of development of the whole, and hence brings them to his own recognition and consciousness. He can consequently understand his undeveloped pupil in the laws of his development. He can stand encouragingly and testingly beside his pupil so much the better as both are beings of one kind, both are human beings, and he (the educator) is conscious of the fact.

The nature and the general laws of development of the whole are expressed in educator and pupil, although in each in a separate way. For this reason misunderstandings and misconceptions will come in with the influencing of the educator and the achieving of the pupil, in spite of honest effort on both sides. The pupil, as well as the educator, is, as a part-whole, a separate being. So the influence which educates by developing and which is as free as possible from errors requires a higher scrutiny lying beyond the changing, separate life and the misconceiving separate nature. But to the thinking educator who has become conscious of this vocation and its requirements Nature is a part-whole, is a touchstone, facing him in his separate existence. A sure means of criticism which is given to him by Nature is its mute, but yet clearly speaking laws of development and formation, which are necessary, not free, but changeless and indeed eternal. A second means of criticism is afforded by the free spirit which has become conscious of itself in its laws of thought, which are likewise eternal, but spiritual. Both the laws of Nature and the spirit's laws of thought reciprocally explain, confirm, and complete one another because they have their common, final cause in the original nature and original life in the eternal and single goodness—in God.

We should be now already in the position to answer the question, What is the touchstone for the kind of training which educates the human being by developing him? Yet in its nature, as well as in its results, this question is so important for the human race, humanity, and the nations that it requires not only earnest examination, but also the provision of a touchstone (or test) which lies within its reach. For as Nature develops in constantly equal, quiet, and inviolable necessity and uniformity, so does humanity develop in continual change from dependence, freedom, attainment of consciousness and independent choice which lead to the discovery of that which is right and necessary. This change is effected by erring and failing, by ignorance and imperfection, by that which has taken place and is taking place to the discovery of that which is right and necessary. But that which has taken place and is taking place (which latter immediately again becomes the former) refers either more to the inner, the mind, the nature, or it refers more to the outward, the relations, and positions. But that which has taken place in this doublesidedness as the history of the inner and of the outer life shows in its results, limitations, and laws of development how it exerts a cultivating influence in and by means of these results while developing itself in accordance with necessary laws and referring to these laws. Therefore we now also see humanity, the human races, and the nations securing the right that the results of history, as well in its separation as inner and outer history as in its union (history as such), should go side by side with and test the education of their children.

And so then it is also possible for us in reference to our subject (the kind of training of the human being which educates by developing) to give to the third and last question (What is the touchstone of the correct following of training?) a complete answer which meets the question on all sides. Firstly, and first of all, Nature in its necessary, changeless laws of development and formation, then the intellect in its unchanging, logical laws of thought, and finally history (the history of the inner and invisible, as well as of the outer life) in its actually manifest results.

The decision of such tests is to be trusted wherever such tests appear, and this so much the more as they coincide in the same decision. We now see at the end of our contemplation what is to be expected of the future in this respect.

Let us now return to the course of training employed by Lina's mother, and let us follow it observantly in order to perceive in it more of the nature and requirement of the training which educates by developing.

Lina wishes to learn to write. The mother complies with the child's wish, knowing what writing is—namely, a memory, a thought (therefore originally purely internal and invisible) afterward as spoken word audible (therefore perceptible, though vanishing on the instant it is perceived) connected with a visible, abiding sign. We now see the mother act in conformity with this knowledge, and proceed from this sure foundation. As has been frequently mentioned, she connects her action with the name of the child. What does she wish to obtain by this connection beyond what we have already noted? First of all, she requires the child to feel and to think of its own personality. She reminds it of its name. She thus requires the child to feel itself to be the precise person named, and to think of itself as such. All this is in reference to the child, and is done by the child inwardly and invisibly. This is the first step. But the child desires to write its name; it desires to connect its internal, invisible personality with that which is outwardly visible, and therefore a pure opposite of the former. This is the second step. What does the mother do to attain this? She lets the child speak its name, which was first thought of inwardly and invisibly, and which it desires to represent outwardly and visibly (to write).

What is speaking? Making an inward thought outwardly perceptible in such a way that it vanishes again the instant it is perceived. Now, what is the word according to this statement? It is that which is intermediate between the purely internal, invisible thought and the completely external abidingly visible sign (the writing). It unites in itself the nature and the properties of both thought and writing, and thus connects them. For the word, being only audible, is invisible, like the thought, but it is, however, outwardly perceptible through one of the senses, that of hearing, as the writing is through that of sight. This is the third step. Lina's mother has found another secret of that training of the human being which educates by developing, and a further sure foundation for the accomplishment of that training as well as for the attainment of a training which meets on all sides the demands of the human being by clear perception, knowledge, and recognition, as well as by the conscious application of the connecting third (between each two things, qualities, etc., which are opposite to, yet like one another). The law of connection is the fundamental law in the universe, the fundamental law of the visible and the invisible, of the spiritual and of the corporeal world. The presentiment of this law was to man the first sign and seal of his nature and of his worth. Man and humanity are the representatives of this law, for man and humanity stand in the universe between God and the creation. They are to recognize both God and the creation, and are able, destined, and called to act as God acts in their life-course in and through the creation, because they themselves are created and creators.

For confirmation of what has been above stated (without appealing further to the severe but just criticism above named), let us follow further the way of training of our much-mentioned mother. She now requires the child (after it has recognized itself as a person and felt itself to be such) to speak its name plainly, and listen to it attentively. Let us also listen to the child as she says, L-i-n-a, and like mother and child we also hear the sounds i—a. "But your name is not formed only of these two sounds, so say it again attentively, and let us see what else it contains." "L-i-n-a," says the child again, and mother and child find that the name contains the articulations L—'n, besides the pure voice sounds (vowels). Indeed, in the course of their way of training the mother and child find that what is spoken also contains such sounds as b and p, d and t, g and k. The mother lets the child feel and anticipate that speech itself, the result of the thought and intellect of man, also contains sharply defined opposites, the tones, or voice sounds (a, o, u, i, etc.); the toneless b, p, d, t, etc.), the so-called close sounds or mutes;

and the continuants (l, n, r, m, etc.), which connect the two former, resembling each other in some respects.

Thus our mother's course of training shows us how man is a connection between the Creator and the creation in the universe. It shows us how language is a connection between man and things, between the thought of man and his action. Thus speech itself, in its first and most exterior elements, impresses the law of connection. Indeed, this law is again expressed in the elements of speech, in the individual parts (to cite but one example out of many possible ones), in the voice sounds themselves, since the sound o connects the two purely opposite sounds a and u, the first of which expresses materiality, and the second the essence. In the national dialects the a like the u passes easily into o. We can not combine the a with the u in the sound of plain "ow" [au in German] without being obliged to use also the sound o, so that when we say au we actually say aou, although the sound o, which is unavoidably used between the others, is but little heard, and still less noticed. Language is an organic construction (Bau) of opposites, a whole which is in itself single, a finished whole.

But language is a result of the thinking mind. Consequently the law of connection is an essential law in the human mind. From the narrow point of view, and within the narrow limits of teaching writing and learning to write, this must here suffice to clearly demonstrate why the law of connection is also the fundamental law of the kind of training of the human being which educates by developing. The individual man, like all humanity, is surrounded by the manifestations of the law. As then he represents this law in many ways, and is, becomes, effects, and creates that which he is, becomes, effects, creates, and does, etc., only by applying and using this law, so a kind of human education which gives such peace, joy, and freedom as satisfy man's nature on all sides is possible only through this law, as has been already said.

As we now see at once, this law is a *practical* one—that is, one which primarily has its determining conditions, its seat, in each essence—but only man himself can become conscious of it in all the directions and references recognized by him. First, by means of this law man rises from the conditions of natural necessity (to which the creatures below him constantly remain subjected) to that of intellectual freedom of will and self-determination. Second, this law, in its true recognition and insight, is not only acknowledged, but even practiced and applied, because it exactly coincides with the nature of man. Indeed, as we already saw and stated, by means of this law man first actually becomes a *human* being—that is, recognizes and acknowledges himself in his essence, nature, and vocation—and in this acknowledgment acts, works, and creates according to this law and so must act, work, and create. In other words, he determines to act freely from himself, as this law requires, without, however, recognizing it in its great generality and as a law of the world, as it were. It is illustrated by the most peculiar phenomena in human life, which hitherto, so far as I know, have not been comprehended, still less acknowledged.

The law of connection has been put into action, especially in the Western countries, and particularly in the German nation. When recognized and acknowledged as a directly practical law (because only in connection is man's existence full of life) it gives to man, as an individual and in communities, that for which (based on his human nature) he yearns, hopes, and strives; since the proofs of it, as well as the conditions of it, are not outside of, but just in the man himself.

By means of the course of training which Lina's mother employed and by the actual inner nature of this training there now lies plainly before us also the nature of our training of the child and man—a training which educates by developing.

We know *what* is to develop in the child and man; it is the godlike in natural, earthly, human manifestations.

We have learned that what is to be developed

by education must be developed by educating in accordance with laws.

We have found that an infallible *way* and *means* of *testing* must be given for the correct application of these laws of education, and we have recognized them in *Nature, in their original cause,* in the *original fount* of all being and life, in man as a thinking being, in the *laws of man's thought and reason*, in the results and *evidences* of the internal and external *history* of the human race and in *history in general*.

We found even at the beginning that the first condition and, consequently, really the first law of all phenomena whether past or present, is the *law of opposites*, and that this law is, as it were, the portion of each being that has entered into existence, (therefore especially of man, who is called to thought and research, to comparison and reflection, to the knowledge of his nature and to consciousness) and conditioned even with man's appearance on earth as a child. Indeed, we find that man, even in his earliest childhood as a manifold part-whole, reveals and demonstrates this law by and in himself.

But we find the *law of connection* given at the same time as the law of opposites. The former is precisely the one which we recognize and maintain to be the inalienable law of all true education, consequently the fundamental law of the kind of training of the human being which educates by developing (our way and manner of educating). Finding it impressed in Nature, and instinctive (that is, determined by the impulse of life) in the life of man, and especially practiced by the mother, we wish to rise to clearly conscious, constant acknowledgment of the fact in life:

First, by means of this law the child is thoroughly comprehended in his nature and in conformity with that nature.

Second, by means of this law man, even as a young child, is recognized and acknowledged to be in the central point of all the relations of life, and the possibility is given to him to fulfill the requirements of these relations.

Third, by means of this law man obtains an evident, sure aim and a satisfactory object of education, and at the same time the suitable medium, ways, and means of attaining this aim and this object.

Fourth, this law and the manner of education founded upon it are, as already above presented, of a purely *practical* nature—that is, pressing at once toward accomplishment and application. Indeed, in reference to life in general, and also in reference to the nature and the requirement of the child as a whole and also a part of a greater whole, and therefore also in reference to the relations surrounding the child and to those surrounding the man.

Fifth, this education (as it presses on all sides directly toward practice and application, and also not only shows, but even gives means, nature, and ways for this practice and application) is directly suited to the time, as well as to the space, and therefore to the space of time—that is, it is wholly suitable for exactly the present relations of the requirements of the present time and to the present stage of cultivation. For our time is a purely practical time—that is, it has now finally come to the point of introducing into life and applying in life that which has been hitherto recognized and which has also been everywhere sufficiently confirmed by experience.

Sixth, like the air and water of life, this method of education (just because it is thoroughly practical, as it exists at the same time with the life and natural relations of the child) also closely connects itself with each age of life, with *each stage* of development and each relation of the life of the child, as well as of his parents and of families, with which latter again it is intimately united.

Seventh, this method of education corresponds wholly by its fundamental law to the requirement of the present time as a time of separation, of isolation, opposition, and contrariety. Indeed, this method of education appears unconditionally (hence required, we might say generated, by itself) as an education of *unification*, and consequently of the actual agreement which is needed for all the relations of life, as well as for the innermost nature and outward existence of the individual human being. For it has indeed proceeded from the knowledge of the *state of being inwardly united*, of *unity*, of the *opposite* (the opposition) and *connection* (as the *first law of life*). This fact is vividly and beautifully and truly expressed in each perfect family, and it is also clearly and precisely expressed in that of our Lina as the *law of the triune life* of father, mother, and child.

Eighth, the *law of the triune life* has been hitherto little understood, but, on the contrary, much misunderstood, and yet it lies directly near to the life of man in many ways. It can only be brought to clearness, to perception, and to the necessity of consciously applying it by the repeated observation and demonstration of the law of connection. The high importance of this law for life in general, and especially for education (the developing education of the individual human being as well as that of humanity), requires that it should be thus brought to clearness, etc. It is also a need of the time and of life to present in life itself the practicality of this law for everyday life.

Ninth, this method of educating is suited to the times, for as it is practical (that is, creating) and closely connected with all the relations of life it develops man even at an early age for a future securing of his subsistence, without early deadening him to a machine among machines, and without diminishing his enjoyment of his childhood and youth. It does so not only because it unfolds, strengthens, and exercises the qualities of the child to the point at which he can maintain himself, but also teaches him to know and treat the necessary material according to its nature, and furnishes and shows the means and ways to do this. It therefore shows and gives (to connect what has been said with a word used at the present time) the means for lessening proletarianism. It restores to *work* its high significance, since it calls forth experience and insight from the creating activity, from the inventive and judiciously accomplished work; requires the cultivation of the *capacity for thinking*, and thought itself; effects the cultivation of the reason together with that of the *power* of will and *action;* lays the true foundation for the training of *character* and of *self-determination* (*in and by means of the life in the social whole*), which is needed by us all and by the world, by each individual human being.

The effect of such education on Lina already shows all this, although only by slight intimations here and there. Its further demonstration, however, by means of our mother's course of training is the problem which we are yet to solve.

Let us therefore continue to observe the mother in the course of training her daughter. She first connected her developing instruction (and consequently education), as has been already brought forward, with the personal feeling, with the observation and recognition of the personality and selfhood of the one to be educated and instructed —of her daughter. The source as well as the object, the issue as well as the aim, coincide in the child. And this, including what are in themselves opposed within a still primitive unit (which we remark in the educating and instructive action of the mother in reference to her daughter Lina), is precisely the *point of germination*, and forms the actual nature of our way of training human beings and guiding children—the method of educating by development. For from this personal feeling (the feeling of self in contrast with the outer world around) proceeds all human education, all education of the *child* as a human child directly upon his entrance into the outward visible world (as a being belonging *to* it, and yet again different from it). This double feeling, this keeping quiet possession of that which is one's own and that which is foreign to one, of the united and the separated, is the germ and, at the same time, the original principle of the germ and root of the training of man upward and outward to a *person*, to a *character*—suffice it to say, to a man in the full meaning of the word. For the perfect (vollkommen *) feel-

---

* Literally fully come.—Tr.

ing of self immediately requires the complete (vollendete *) feeling of the all.

Through this observation of the course of training which Lina's mother followed, and the conclusions derived from it, we have now found what we missed in the education of man up to this time —as entered into in clear consciousness, and carried out and accomplished with clear consciousness—an unchangeable fixed point of union and starting point for all education. This point, bearing within itself the fundamental laws of all education, continues to develop and cultivate itself by this law in accordance with the laws of life, organically in and *with* all, *from* all, and *by* all which takes place with it, *from* it, and *for* man. So developing, this point of union and starting point resembles a seed, the innermost part bearing within itself the *whole* tree, which develops constantly and in accordance with its own laws and those of Nature during hundreds and thousands of years.

This eternally sure and fixed, as well as clearly defined starting point now actually constitutes the nature of our way of training human beings and guiding children—the way of educating by developing—as such a way of training is only possible through this starting point. For it can only appear as a whole in consequence of the fact that the

---

* Literally fully ended.—Tr.

limitation of plurality, multitude, manifoldness, universality, already lies in unity as such, and that an outward and *the* outward is at the same time given in and with the inward. Unity and universality, inward and outward, are pure opposites; but, as a creation becomes possible only through the intimate union of both opposites, so only by that intimate union of both becomes possible the continuous development of all creatures, and, above all, the clearly conscious developing, educational, training (in accordance with outlet, purpose, and aim) of man as an individual, as well as of the human race and of humanity (again as a whole which is in itself single). All that is required for perfection and completion by the education and training so often mentioned is at once given by this intimate union of opposites from which it necessarily proceeds. So also does the further instructive treatment of the mother, guided by which (as by a red thread) we now continue to advance in the presentation of our way of educating and training.

The mother does not let the general perception given by the union of opposites continue to be only a dim feeling; but, as has been already several times mentioned, makes the perception objective to the child by its name. Yet, as was already said in the beginning, the name belongs to the complete realization of the individual as a person and as a character—to the education of man. It also again directly and intimately unites the opposites in itself; for as it separates (isolates) man, so just in and by means of this isolation it places him again in the center of the great whole of life, and, first of all, in the center of the lesser whole of men and nations. The name makes the man able to turn toward each human being while retaining his personality. So again each individual human being, as well as all human beings, can again turn to him. Hence the name, first of all, lies exactly between the individual and the universal. Therefore the stage of development and the epoch of humanity now newly begun lays such great stress on the name man. "Be a man!" is the starting point, as well as the goal, of the demand of the present time. Hence the training of the German people and the national training lays such high importance on the consciousness of being "German"—"a German." "Be a genuine German" is now the principal demand in each German-born individual, whether child or adult, as is the above demand, "Be a man," upon each human-born individual. "Impress complete *humanity* upon the *German characteristics!*" is therefore the uniting demand of education—a demand which is now addressed to us with such manifold clearness, and which it is so indispensable that we should readily obey if we would not deny ourselves to be men and Germans —German men. To attain to this ready obedience and to avoid that denial is the principal effort, the aim and the object of the kind of training of the human being which educates by developing (a kind of training which we introduced). Such is also the aim of that which proceeded from and is founded upon this training with equally inevitable necessity—the kindergarten.

With a deep, motherly, natural impulse the mother has felt through all this, and (by following the instinctive motherly impulse) has at the same time presented the starting point, the source of our way of training which develops by educating, and has made it known, so that it may be scrutinizingly observed, that life may be in accordance with it, and that it may be represented.

All that has durability and firmness bears a reference to a certain internal or external point, and, as it were, rests on this point. This is especially the case with all that is to show existence by life and continuous development. All such must bear within itself a point of vigorous life. Lina's mother's course of training has plainly demonstrated to us that the genuine education of the human being in childhood, as well as in later years —an education abounding in results—must likewise proceed from a fixed, precise, healthy *point of germination*, and indeed from the point of complete union—the opposites—which reciprocally so penetrate one another and so coincide with one another that even the gaze of the innermost spiritual eye perceives no difference, and which again freshly sprouts forth from this point of union, as a point of *life*.

And that training of the human being which educates by developing also recognizes this point, just as bearing within itself the conditions and fundamental laws of the appearance of life, of the disclosure and manifestation of life, and the onward course of life from it in accordance with its nature as the starting point and source of all genuine education, such as leads to the aim and attains the object. For it can proceed (free from doubt and arbitrariness) only from a point which bears within itself at the same time the limitations, cause, and laws of all following development appearances and requirements. Also education, if it is to rest on a firm foundation, must be subjected to the fundamental law of life and existence, of Nature and of the world. It must start from a precise point from which (as existing in and at the same time with it, in accordance with fixed and sure laws of development and formation in which necessity and freedom appear to have an equal right) all the rest proceed in reciprocal balance.

And thus (as the training which educates by developing appears thereby to be the only one corresponding to the nature of man and to child nature) is also gained the first changeless base, the sure starting point, and the pure source from which, like a clear stream, the further developing laws of education—the laws of opposites, of the part-whole of life, of connection, of triune life—quietly flow forth (neither disturbing nor clouding the others), all of which Lina's mother's course of training shows us perceptibly and actually. For —and this is the further highly important fact which lies in the background of the mother's whole style of management, which starts from the instinctive and rises to clear consciousness, as well as to clear insight, and from this manner of management also definitely speaks—all these laws, requirements, and conditions form actually one and the same law (though they appear different and are perceived and comprehended as different)—namely, the law of the original unit—of the being and life which has its source in Him who by himself, in himself, through and from himself is good—God. This law declares and reveals itself as divine in the whole and in each individual being of the all, as the creation of God. This law, above all, lies in the blossom and fruit disclosed before our eyes, in the man who is to be consciously educated to consciousness, and, with and by means of natural necessity, is to be educated to freedom. By this law are given for the subject the " divineness of the nature of man " (as the object of that training of the human being which educates by developing), at once clearly and with precision, the stated laws, conditions, and requirements of its development and cultivation. But quite pre-eminently the infallible means and ways of testing are also given, that they may be employed. These means are: *Nature*, the laws of decay, existence, development, and formation of and in the universe, in the creation; the *spirit* (in its eternal laws of thought and in accordance with them) and that which is recognized in it and by means of it, namely, *history* in its results and laws of life perceived as separate, in the history of the outward and of the inward life, or as history of the whole, joint, united and single, inward and outward life. By means of all this, and this is shown to us even by Lina's mother's way of management, education has again become (or is now actually for the first time by means of that training of the human being which educates by developing) what it should be—a *science*, a genuine *science of education*, an education with clear knowledge of the subject, of its aim and purpose, of the means and ways, etc. It becomes an *art*, a genuine and true *art of education*, dependent on a vivid, all-comprising *idea* of education. It is, above all, a simple, practical (that is, easily and clearly practiced and practicing) *living fact*, which grows forth to a genuine life of education leading toward the aim—the educational life of the individual, of the family, and of the people—which rises from the instinctive impulse of Nature and life, an impulse which (as a true daughter of Nature) leads us to virtue, etc., to self-stability, morality, union with God. By such an education life in all relations and endeavors is satisfied.

It only remains for us to point out in general how all this also is realized and obtained by that training of the human being which educates by developing.

The question here is, first of all, practical understanding. Without entering further into the division of education just indicated for the clearer comprehension of and the deeper search into the subject—viz., the division into the life of education, the art of education, and the science of education—let us now turn rather to education as a finished whole, single in itself, pervaded by life, art, and science in equal measure, as, in the just-mentioned arbitrary and artificial separation itself, either may be predominant by catching the eye before the other. In order to solve our before-mentioned problem, we will now show how we apply the laws, conditions, and requirements recognized and explained in the preceding pages in the representation of that training of human beings and children which educates purely by developing, always with an explanatory retrospective glance at the way in which Lina is comprehended by her mother in the course of training pursued by the latter. As we now consider our child, the little one whom we are to cultivate and train, as an individual and particular thing which is conditioned and demanded by the whole and general, and which bears within itself the limitation of its existence, of its development and cultivation in reciprocal action with the life-whole, we look upon each of its expressions of life and activity as a purely personal expression of its own life, but (even in the smallest of its expressions) constantly in combination with the great whole of life and with the All-life, standing in relation to it, in two respects—once in reference to its outward *appearance* and *effect*, its influence, and again in reference to its *inner* origin, to its original source in its own nature.

Helpless, indeterminate, and weak as the child seems in all his expressions of life from his first entrance into life, he does not perform a single action which is to be isolated, and is not to bear within itself *at the same time* the *three* relations of individual and *personal* life, of life in the *whole* (in Nature), and of the *united* nature of both; consequently there is actually no action which is not a *triune* one. For all expressions refer constantly to his personal existence in conflict (of comprehending and doing) with the outer world, a conflict which is mediated and removed by the spiritual union of both.

So now also our way of training, which educates by developing, comprises from birth each phenomenon in the life of the child—first, for the securing of his existence as a personal and separate being; second, for grasping and handling, for understanding the outside world around; and third, for the arousing and fostering of the presentiment of an *individual* and uniting nature. All three of these have been done hitherto even by each simple mother guided by her natural impulse as a human mother, so especially—which perhaps seems wholly unfounded and strained to many— by her talk to and with the child from the first instant of his claiming and appropriating, on through childhood. What has hitherto been done (and always, even by the mother) as a natural impulse we now raise to action with clear *consciousness* and true *insight* and *circumspection*.

By this comprehension of the child, by definitely bringing out this comprehension (which is needed) and by placing it in a clear light, all which is done and is to be done by and with the child receives its true significance, even the bodily tending, the providing of food, and the motherly petting. This view and treatment also blend with all that is done with Lina and with the way in which it was done. That first attention to the child in its triune life by bodily tending, by food, by the offering of nourishment and by motherly petting, is connected with the development of the limbs, senses, and mind of the child as a triune being. For, as soon as anything particular is brought near the child's eye, which arouses his innermost life and will, we see immediately that the development of the senses acts on the thinking powers, and at the same time on the will, the use of the limbs, and the bodily activity. How hand and foot strive for suitable activity and right use!

The child between the ages of six and nine months strives already for the free use of his members—first of all, for that of his arms and hands. But since now the child is an observing and imitating being, we see how he when not yet nine months old imitates little movements with his hands (turning the hands, clapping, moving the fingers). But as this imitation is by no means merely mechanical, a merely external copying, as it were, it is evident that the promoting of the child's play by the mother, her talking to him, her entreating wishes, are essentially effective. We recognize from this statement (connected by language) how with the slightest definite activity of the child's limbs his power of thought and the power of his senses are also active. We see here again three activities united in one, and we also see in general the alluring charm, the retroactive impulse, and the comparing activity (which three form an action in itself single).

The influence of the word is yet more heightened by the law of movement (the rhythm) and by the singing tone (the mother's way of singing), because, in this way, the word has an influence on the mind, on the thought, by means of the feeling.

The early, harmonious, joint comprehension and this treatment of the child which educates by developing find in the Mother-Play and Nursery Songs their living expression and actual production, which are proportioned to the different stages of childhood, and at the same time explain and point out the inward spirit in the outward appearance. The Mother-Play and Nursery Songs proceeded directly from my observing the actual life of mother and child. The understanding of this book was therefore supposed to be easy and the work was committed to family life without introductory words. But as often as the life of the mother and child and the reciprocal life of both is repeated with each newborn child in each family experience has shown that the life of the child in relation to the whole family through all conditions of life is, alas, only too little observed. Hence the Mother-Play and Nursery Songs, just named (although a pure demonstration and necessary continued development of actual life), has been hitherto so little understood, so little acknowledged, and still less brought into the family and used there. After our own diversified use of it for many years, and especially after it has been used and tested in many ways by thoughtful mothers, we must, without regard to its authorship, recognize and acknowledge that it in fact not only indicates the actual starting point and source of the true conscious training of children and human beings which educates by developing—necessarily required by the present stage of the cultivation of man—that it not only shows the means, way, and manner, the object and aim, of such training, but also actually produces, at its most important period a family life which fosters childhood in such a way.

How Lina Learns to Read is a continued development of the earliest observation and management of child nature. May this chapter, in connection with what has been before stated and demonstrated, serve as a test of the Mother-Play and Nursery Songs by means of the observing of the life of mother and child separately, as well as in connection with one another in its inner foundation, in its fostering, and explaining the actual life of the child and family, and in its effect upon and result in the education of childhood as well as of man in general. Such a test would at least aid in the true understanding of the above-named book, and also in the thoughtful use of it in the family. For through the comprehension of the training of the human being which educates by developing, and which is founded on and presented in the above-named book, the child enters into his right relation to himself as a separate being enters into the surrounding world as a part of it and (by the help of language) to the uniting and single spirit which lives in all, as is presented by the whole course of the guidance and treatment of Lina.

II.

THE CHILD'S DESIRE FOR SIGNS.—INTRODUCTION.—MAN A CREATIVE BEING.—A CREATOR IN HIS SPHERE OF ACTION.—EARLY DEVELOPMENT OF THE CREATIVE POWER IN THE CHILD.

WE recognize the fact that man, especially in early childhood, is in intimate, united interdependence with Nature and its course of development—a course which is in accordance with manifest law. But in order that the course of his developing education may be assured and clear (therefore for his own welfare), it is at least not less important to regard man, even as a child, as respects his nature and activity, in the most intimate and lively connection with the Original Cause, the Creator of all things, and with the oneness of his creative nature.

To view the child as united to Nature gives security, conformity to law, recognition and insight, firmness, applicability, dexterity, and extent to the education of the child. To view him as united with God gives dignity, truth, clearness, light, infinity and unity, spiritualization, sanctification, blessing and blessedness to that education.

But in what way did and does the eternal, Original Cause—the Creator of all things—make himself known to us? In what way does he reveal and manifest his nature? Just by his eternal action; by his eternal, uninterrupted creating; by bringing into existence from the eternal spring and fount of his own being; by the manifestation of the invisible oneness of his being in the visible appearance of the individual; by the endless revelation of his own nature, which is in itself one, in the innumerable manifoldness of individual existence.

And now by what means does man, even as a child, make known his being in that which is phenomenal? his nature in his existence? How does he make himself known, and thereby cause us the purest joy (and, in the course of development, entertain and even astonish us)? Is it not by action, by activity? Indeed, when the use of the child's senses is but partially developed we must recognize the activity of the child to be comparatively observant, usually, indeed, excited from without, but yet actually and finally determined by the innermost workings of the soul (therefore, as it were, created from the invisible spring and fount of the soul) to create, therefore to employ himself, as one called forth from hidden being into existence, into perceptibility.

In this steadfast contemplation, and with this view of life, let us now observe the voluntary and spontaneous expressions of life in the child scarcely a month old in regard to the ultimate innermost and hidden, constantly invisible, original cause of these expressions. Let us see how the child, with the united power of soul and intellect, strives to call into existence for himself and, as it were, from himself, that which does not yet exist for him, which is not yet in the province of his perception and recognition. The child by this striving shows himself to be a creator in his own little world. If we thus perceive and experience, we must recognize (being forced to the recognition by the perception, experience, and actuality of what we have observed) what indeed shows itself already as the result of pure thought: viz., that man, even in childhood, proves himself by his creative activity (which is conditioned in the innermost parts of his being, as it originates in the inscrutable Eternal, in the Original Cause and Creator of all things) to be like his Original Cause in that he is a creating, creative being. And thus man, in accordance with his nature, in and by this creating (which shows itself in the child as an employment of self) shows himself to be related to his Creator?

But now in what single phenomena, in what manner, and by what means does this strengthening and elevating, invigorating and encouraging, purifying and blessing, even hallowing relationship of the child to God as his Creator, and consequently as his Father, make itself more fully known?

The genuinely healthy child will be always active, he will employ himself. Why? He wishes to make something so that his inward desire may also appear externally. He wishes that what is hidden within him, and lives in him, may also outwardly exist. Therefore as the inner conceptions, the intellectual perceptions and comprehensions, the images of the soul, change in the child, so also the activities of his life, which are taking form, change with equal quickness.

But now what is the further cause of all activity in the child? It is just life, as life is the first cause of all existence in God. Therefore the child invests with life whatever he sees—that is, he not only anticipates, feels, and experiences life in all, but he even attributes conscious life, will power, conscious, self-determining will power to all. As all things emanate from the self-determining will of God (the First Cause of all things), and as life and that which has life and which veils life proceeded from God only, so the child sees and anticipates hidden life, and that which has life in all his surroundings. This fact shows definitely the relationship of the child in his activity to his First Cause and Father, as the Creator of all things.

Yet the proofs founded on facts of the child's inward relationship rise ever higher. The child not only anticipates and imagines life in the objects around him as soon as he places them in reference to himself, and himself in reference to them, but all which proceeds from the outward and inward life of the child appears to him immediately in completely finished animate form. Thus in the things he makes or with which he plays he sees the kitty, the birdie, the little fish, etc. The lambs are represented now by white beans, now by the flower buds of the field or pasture. Sticks must realize the idea of trees. Blocks, etc., must be persons. Indeed, even the child's own fingers must spread themselves out and become now different children, now little fishes, now little birdies, etc. Thus the child, whose life is a whole in itself, at first always represents life as a whole in the objects around him, since each and every thing which has entered into existence from the being, life, and action of God the Creator is a whole, and is also a part of the great All-life.

Not until a later period, when the examining power of his reason is more developed and his creative and creating power has become more independent as well as spontaneous, does the child compare and separate.

Now, in what sequence and in what way does the child early reveal and manifest his impulse to activity, to employment, and to representation? How does he reveal his operative, creative power?

The first object of the tendency to activity and employment, the actual attraction for forming, is the child's own limbs, often its whole body. He joins his little fingers and hands in different positions, and even seeks to represent different objects by them, as also by his whole body. This is, as it were, the first development and preparation of the limbs and body for creating representations by other objects and materials.

These other objects are, primarily, solid, bulky, capable of being grasped by the hand, firm. They are at first tested by the child as to their power of standing alone, their movableness, their pliancy, their capacity for being united, the possibility of easily joining them together and again dividing them from one another. Spheres, wooden blocks, stones, the ball, are therefore the first playthings of children.

By using them the child will produce outside of himself that which he conceives within himself. This is a proof of his tendency to do something, to produce (his creative impulse), and a token by which he shows this impulse. Therefore, even the indication of the child's activity is important, but his later efforts to draw are yet more so.

> Ah! this little child, I see,
> Would e'en now an artist be.

Perhaps with the second, certainly the third year, this bulky solid form material is replaced by other materials of two different kinds. The one is yet more bulky, but can be easily impressed. It consists of soft loam and wet sand, water itself in its movableness, its tractability, and capability of being guided, and the air with its power of moving and turning. The other consists of less bulky and solid objects: small flat pieces of wood, little slats, smooth paper, or sticks and thread. Finally, the child chooses dry sand, sawdust, plate glass moistened or breathed upon. He also chooses objects which by friction leave a mark, such as slate and slate pencil, paper, lead pencils, colored chalk or crayons, and colored liquids—that is, the colors themselves. Hence the child's desire for drawing and painting. Both are quite essential, developing means of education and cultivation of the child and man. But singing is no less essential. For even the easily resulting and again easily vanishing, echoing tone produced in one's own throat or by one's own members, or by ringing and resonant objects (glass, bell, metal, etc.), must serve for creative representations of inner conceptions, sensations, feelings, and indeed ideas.

Thus we see how an advance is shown in the child's creative representations as the means of play lessen in materiality. The forms made with the solid material often give but a slight outward representation of an object, but were mostly called forth by fancy. The forms produced by soft material showed more the inward connection by the outward form. The sticks rudely represented the outlines. These appeared more complete in the sand and dust, as well as on the pane of glass which has been dimmed by the breath when the forms have been made by the easily movable finger; but they are yet more sure, precise, and complete, but less material, when slate pencil, lead pencil, etc., are used on slate and paper. Yet the echoing tone in its harmonious and rhythmical, as well as in its melodious combinations, expresses directly the higher and the highest feeling of life in its unity, one part flowing into another. It is the soul which here speaks to the soul; the life which speaks directly to the life through the life (especially connected with the composite, immaterial word), whereby the spirit speaks to the spirit. But in the spirit and by the spirit man recognizes himself as a creative being; he recognizes God as the Creator, and he recognizes Nature as that which is created from God. Thus we see and recognize that we, by fostering the creative power in the child early in his life and through the stages of development indicated by the child himself, raise him to knowledge of himself, of Nature, and of God, and to the recognition of himself as a child of God; but " by their fruits ye shall know them," says the highest educator of humanity. So we must recognize here that we, by early, continuously, and symmetrically developing and cultivating man's creating power in conformity to law, raise him to the true dignity of human nature, to fitness for life, to accordance with Nature, to genuine all-sided union of life, consequently union with God—therefore to true peace, to pure joy, and to constant freedom.

The Child's Desire for Typical Representation.

The object of the previous essay is to lead us to observe the child and the child-world, and to perceive in both the truth that man is a creative being. If we now look back again upon both, we see that the child's activity, taken collectively, from his first spontaneous movement to the stage at which he has gained the power to make a life of representation, of the life of feeling and sensation (that is, up to the age of completed childhood, therefore up to his sixth or seventh year), has its foundation in the effort, first of all, to make known first his inner life in and by means of outward phenomena as soon as it comes to his perception, to place this life objectively before himself and externally to himself; and next to appropriate the inner life of things around him (that is, to bring himself to knowledge of, and thus to insight into this inner life, in order to reproduce it spontaneously), and indeed to come to a knowledge of it by this reproduction.

In a twofold direction, indeed, but in a self-determining way, which is in itself single, the child's action here depends always upon the comprehension and manifestation of the inner in and by means of the outer. It depends, as it were, upon a creation which emanates from the inner being and shows itself in that which is present and apparent, therefore upon an actual creation; for the spirit, the life, hereby acts as the determining power and conditions the material.

Yet as certainly as this uninterrupted self-revealing, creating activity of the child is in its natural healthy condition a general one, so certain is it also that the power of the child to exercise this activity is still very weak and slight. But that he may not feel himself restrained from using his power by perceiving its weakness, the child who is undisturbed in his development always feels that his power is at least great enough to accomplish that for which he strives. Therefore (as every one who has watched the impulses of healthy children will have been convinced) the as yet slight power of the child is not in a condition to obstruct his impulse to creative activity, but, on the contrary, he seeks to strengthen and elevate this impulse by increasing demands on the efficiency of his power. We must not disturb the child in this effort, though it be often apparently fruitless. If he does not actually accomplish anything outwardly, yet his inner power of creation grows by his efforts. But, as a rule, the child will himself seek out material by the use of which he can gratify his impulse to represent by creating, or, in other words, his impulse to creative formation. These materials, by the aid of steadfast will, will finally submit to the influence of the yet unpracticed arm and the little hand, as we have already shown.

It is now just as indispensable that those who are around him should, by promoting and fostering it, meet this effort and impulse of the child and his activity as it is essential that the moist warmth of the earth should offer to the germinating kernel and the clear, shining light of the sun to the bud which is striving to unfold, the right conditions for the complete development of their powers.

Now what Nature, the mother of all, gives to her children that they may reveal what is within them, the conscious love of parents must supply for their children, and the love and insight of adults must provide for the children of the existing generation for the free development of their nature. "Draw a mousie for me or a little house," "Paint a birdie or a flower for me," is therefore likewise one of the first requests of the child as soon as he can make known by words his will and his inward impulse. He also begs, " Do tell me a little story," or definitely, " Do tell me the little story about the birdies who loved their mother." By this means the impulse to representation and the

DRAWING THE MOST IMPORTANT AGENCY. 67

power of creation grow in the child. Now, as soon as he can master any kind of suitable, plastic, flexible material he tries to show his impulse and his power by representing, forming, and creating—by employing himself in manifold ways.

Now, although all that the child does is a creating from himself (even his plays with the most palpable, most material substances—cubes, blocks, pebbles, etc.—being a kind of painting or drawing of his inner self; that is, of that which lives within him), yet it is painting and drawing in a narrower sense, even if it be only the drawing in the earth and on the pane of glass moistened by the breath, which has been previously mentioned, that attract the child above all and ever anew as a means of representation of his inner self. But why? Because this gives to the operative impulse to formation and effort in the child an all-embracing satisfaction; for the child can by the drawing just as well represent a star as it shines in the sky as the flower which blossoms in his little flower bed. He can thereby just as well represent a tree showing itself in the woods as the flying birdie sitting on a tree or fluttering its wings and rising into the air.

But this requirement of the child to avail himself of the most easily movable, the finest and smallest material for his producing and drawing, for the showing of his little creations, for the manifestation of his power of creating, is now fully in harmony with the doings of Nature and with the phenomena of Nature. For Nature also creates her works from, and represents them by, the most easily movable materials—light, air, water, earth, dust. And so the desire and will of the child are again shown to be neither individual nor yet isolated, but to be a necessarily postulated living whole in itself single, which is creating and has created, by which showing the child proves himself to be a part of this whole.

Hence we see that even from this point of view the efforts as well as the desire of the child to prove himself, by the aid of the objects mentioned, to be a representing, forming being must be sacred to us. For, as the child proves himself in this way to be a creative being, he also shows himself just as surely, on the other hand, to be a member of the great living whole and of all life. He is destined to develop himself as a creating and as a created being, in and by means of the great living whole, in order thus to have knowledge of the Creator and the creation (recognizing the Creator and understanding Nature), and therefore to comprehend and to bring himself to consciousness of his own nature, since by what he does he stands intermediately between the two. He is therefore destined, like Nature, and like the Creator of both Nature and himself, to create the great from the small, and by means of the small in constant coherence with the universal life, to effect good, to form the beautiful, to show the true, and to do the right.

If now all these activities of the child previously mentioned, and the different materials used by him, admit of this mode of contemplation, and show him as creating (with which doubtless later and further presentations of the nature of the child as a creative being will be connected), yet it is above all the art of drawing by which the child in his circle already proves himself to be a creating being, because with the slightest mastery of the material and with the exertion of the smallest amount of physical power, there can most easily and quickly be shown recognizably by the drawing that which the child would like to represent from himself, that which he would like to create. Therefore now the development of the power of drawing in the child belongs to one of the most essential members of the educational training which develops the human being and is one of the most essential bases of the general education of humanity, of the education of the human race toward union of life on all sides. Such an education has long been dimly anticipated by humanity, and so is now longingly expected.

Owing to the fact that the power of drawing has not been completely recognized hitherto, and that the introduction and practice of drawing has not been generally considered to be an essential part of genuine human training and has not been received as an essential means of educational training, humanity, especially in childhood and youth, has up to this time been cut off from one of its most comprehensive means of training.

Slight as the necessary expenditure of power in drawing seems to be, yet drawing in its application and execution makes a demand upon the whole human being, consequently on the child in all the references of his development and training. Even the correct position of the drawing fingers and hand for spontaneous use requires a correct, suitably free position of the whole right arm; this again indispensably requires a corresponding position of the other limbs and of the whole body of the child who is drawing if he wishes to represent what he creates with freedom of bodily action as well as with a free spirit. For a freely active, skilled use of the body necessarily presupposes a free, skilled spirit in the circuit of that activity. The two condition one another reciprocally.

As therefore true, free, beautiful drawing requires that the limbs and body be symmetrically developed, it also demands the spontaneous, skilled use of the senses, and it no less demands the sense of hearing and that of feeling than that of sight. This wholly satisfactory training of body, limbs, and senses, and consequently the development required for drawing, conditions in the same way a harmoniously unfolded soul, a feeling, experiencing mind, as well as a thoughtfully comparing, intelligent, and perceptive intellect, formed judgment, correct conclusion, and so, finally, an idea (more or less clear, at least more and more improving during the representing activity) of that which is to be formed.

But, again, this demands from, and forms in the child who is drawing observation and attention, the comprehension of the whole, recollection and memory, the gift of connection and invention, fancy. In general, it enters on the path of corresponding use of man's total power of formation, enriching the spirit with clear conceptions, the mind with true thoughts, and the soul with beautiful ideas—the said conceptions, thoughts, and ideas being the fundamental conditions of creating the animate and active. For such creating the child already yearns and strives.

The drawing which, to the injury of the children, has been hitherto neglected in their early education, is of general, universal, and comprehensive importance in the training of the human being. As a complete presentation of his creative power, it renders it possible for man, by the strong impression of pure humanity, to become within himself, and by his own action, a second creator of himself, as well as a creator and outward representer of pure humanity and human nature. Drawing also makes it possible for man to rise from the correct comprehension and cultivation of that which is corporeal, material, and sentient, through sense, modesty, and morality, to true union with himself, as well as with Nature, with humanity, and with God in feeling, thought, desire, and action.

This must here suffice to lead to the recognition of drawing as an essential means of educating man up to completeness—his constant vocation on this earth—conformably to the unity and universality of his nature and the use of his creative power as an individual being and also as a member of humanity.

We have just given prominence to the fact that the recognized goal can not be reached by one-sided development, but only by symmetrical development of both body and spirit.

Above all, we must let the child be early interpenetrated by the feeling that a free, sure, firm, position of the whole body not only makes possible a free, easy use of all his limbs and senses, but renders possible such a use at the same time with a pleasant feeling of tranquillity, whether he sits or stands.

In general, all in which the child's active will is required and necessary must at first be attended by a pure feeling of pleasantness (through as pure a feeling of well-being as that with which the child clings to the mother's breast and is pressed to her heart). With this feeling of pleasantness must be connected at first the becoming accustomed to the right, and later on also the feeling of the right and correct. This feeling also awakens early in the child, often gradually, often strongly, but always easily. This also, like the feeling of pleasantness, exerts a determining influence upon what he does, and upon the manner in which he acts. The development and cultivation of the child to a creating being, even in the special cultivation of the creative power in and by means of the drawing, must therefore proceed from the careful fostering of these two feelings (the one of which appears more bodily and sentient, the other more intellectual and spiritual) early arousing in the child or at least soon to awaken.

Here now with the firm holding and free position of the body begins the cultivation of the arms, hands, and fingers. This was formerly done, in general, under the guidance of the Mother-Play and Nursery Songs, and is now carried on with the special object of cultivating the above-named members for drawing as a creating power and activity. This cultivation of arms, hands, and fingers goes on in rest as well as in movement. This movement is in straight as well as in curved lines, and in all directions. The drawing is at first wholly in free space. Later, it is done so that the traces of the movement (especially if it be a continuous one) may be made perceptible upon the surface—for example, on the earth, in sand, in dust, or in fine sawdust, which has been spread on a suitable plane; and yet later by the use of objects, such as chalk, slate pencil, lead pencil, etc., which leave the traces of the movement as lines on blackboard, slate, or paper.

This last appearance is now (if the consciousness of the precise object be reached) the drawing (showing) of the lines, first of all the curved, and afterward the straight lines, with the practice of which, therefore, the development and cultivation of the child's impulse, capacity, and talent for creating drawing must begin.

Thus the comprehension and representation—that is, the drawing—of the curved and straight lines are not only closely connected with the simple movements of the limbs, but proceed directly from these movements combined with consciousness of the purpose. Both the curved and straight lines appear in different positions. The latter appear as vertical and horizontal, and as oblique or diagonal lines.

Yet, as we recognized, the drawing, being the complete creating activity of man, must proceed through the attainment of consciousness and must be accompanied by consciousness. But, again, both consciousness and its attainment begin with speech and proceed from words. Therefore the showing (pointing out) and the testifying (awakening consciousness) word must be connected with the drawing. But since here also the activity of the child, as always, proceeds through feeling, or, if the other mode of expression be preferred, through the feelings of the pleasant, the beautiful, and the right, etc., the growing activity of the child should not only be accompanied by spoken, but by sung words, thereby leading to the right and beautiful.

We therefore connect the drawing of round (curved) as well as straight lines with the explanatory word or with the animating little song (for instance, the ball and sphere songs before mentioned), in order not merely to awaken but to foster and strengthen the whole, collective activity of the child, as will soon be stated. While the pencil moves in a circling manner on the slate these words, for example, can be sung:

> Around, around; how much I enjoy it!
> My pencil I turn; thus I like to employ it.
> Thou, too, must enjoy it.

Or:

> Do see the straight, straight line
> My pencil makes so fine.

With the round, as well as with the straight lines, besides the position and direction, the manner of origination or forming should be considered in reference to the one who is drawing—from the hand, to the hand; or outward, inward; up, down;

down, up; or the opposite originations may be united in a zigzag or winding course; this, as experience shows, gives great pleasure to the children, especially if to the explaining word, which speaks to the intellect of the child, be added the loving tone, which speaks to his heart, and so as the flowers blossom with the sun's rays which shine upon and warm them in the morning, there result here the many kinds of combinations of lines by the help of the clearing and pleasing words of song:

> Zic, zac, zic, zac,
>   Goes my pencil fleet;
> Tic, tac, tic, tac,
>   Sounds the round clock's beat.

Or, with winding curved lines:

> So the line winds along
> With a song, with a song,
> And the time seems not long.

Yet soon these lines, the drawing of which is now his object, become to the child, who is guided by them to thoughtful notice of that which surrounds him, again a means of further representations—that is, material for representation. Thus, for example, the circular lines which the child can now draw with considerable facility become to him the image of the moon, the sun, a target, even an apple, a ball, a sphere, a hoop, a ring, etc. He has seen in the meadow, in the garden, and in the field the three-leaved clover, with its rounded single leaves, and the five-leaved flowers of the most different kinds, with their petals set round in a circle, and he easily represents them by winding circular lines (as well as rayed flowers and the many kinds of feathered leaves which are often quite rounded; for instance, the pinnated leaves of the creeping rosebush, a kind of field rose, the acacia, etc., or well-paired cauline leaves, as, for example, in the beautiful sunny blossoms of the moneywort). But the child's impulse to represent by drawing ventures also upon the animate. He tries to represent the cony, with its rounded form, the mouse, the lamb, the dove, etc. The child has exercised himself essentially by means of his round plaything, in the clear, sure perception and the representation of that which is round in form.

In the same way that he was attracted by Nature the child is also attracted by the human being, by human life, and so by his fostering place—the house! "Draw me a little house!" We have already given prominence to this request of the child. He now tries to fulfill this wish himself. Now new and differing demands are made upon the child: First, the more acute perception of the different positions and directions, especially of the right lines as defining the positions of the oblique lines; then the relation of the single lines as parts of a whole to a uniting and limiting middle point, line, etc.; finally, and lastly, the exact perception of the relative length of the lines by comparison with a fixed, defined measure.

These three indispensable requirements (position or direction, size, and drawing point) for correct and beautifully formed drawings indicate that the eye should be cultivated as a measurer. The solution of these requirements must therefore proceed from the capacity of the eye for correct perception (as did the representation and execution of the line from the training of all the joints of the arm) as well as from the cultivation of the whole body.

But the drawing in the network is the ultimate reference which receives and, as it were, forms the outer world in the eye. Name and thing show us here in a remarkable manner, and, by the child himself, the way to cultivate the child's eye, and thus to cultivate his sense of the perception and representation of the comparatively correct positions, sizes, and uniting middle. This is a netted surface given to the child, an exteriorly placed net on which he can at first with certainty represent its lines as the condition of bounded surface formation, and, first of all, the straight lines in various positions and lengths, and can thus bring himself to consciousness (through simple, continuously progressing multiplication of a line which is comparatively the smallest and serves as a measure—the line of the first single length). Later proceed from

MAN A CREATIVE BEING.

this the sharper perception and conscious correct representation of the curved lines, the circle again being first.

This multiplication of the first single measuring length (relatively the smallest), applied first of all to straight lines, concludes with lines of five times the length of the first line, which gives the measure. This number is determined for the child by the number of fingers on each hand.

For the purpose of presenting such a netted surface, which will, first of all, cultivate the eye of the drawing child for the perception and representation of the relations of direction, position, and size of the lines, it is best to use a smooth slate, on one side of which a network has been formed by cutting vertical and horizontal lines at always the same distance of the quarter inch with the inverted sharp point of a knife.

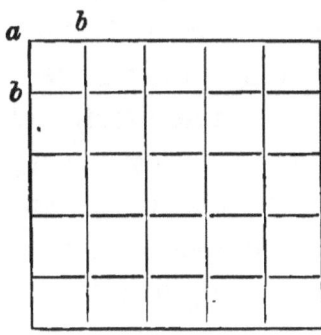

The distance from one of the parallel lines to the next is indicated and measured by a right line of single length or size, and so progressively the distance of the fifth line from the next is a right line of fivefold length or size.

As now the child's consciousness of his power of presenting things by drawing generally begins with the perception and representation of the straight line, so the perception and representation (that is, the drawing) of straight lines begins with the vertical line, and all the relations of size and grouping are first carried on with the vertical lines before the child advances to horizontal and oblique lines.

Thus the child should first draw many times a vertical line the length of one square, then of two squares, and so on up to five, always at equal distances, each time separating the rows from each other by a distance of two horizontal lines.

The drawing and correct measuring of the lines is now done by the child at first in the marked lines in the required position and extension. In this way and by this use the net serves to train at the same time the child's eye, hand, and fingers. But not this only, for the inner law (which is manifoldly expressed in this net as an outward law) trains the thought and spirit of the child, since by means of it he is guided directly at the instant of action to a sharp and comparing conception of what he is doing. For, as already said, the "aim" is not only the production or representation as such, but we aim to make the child conscious of what he has done and how he has done it. Thus the child's doing becomes a creating, and his activity an individual employment. This is done when the word, which makes the action objective and leads to the child's attainment of consciousness, is joined directly to the action, and the accordance of word and deed constantly and strongly expressed by the silently compelling law in the net, thus leading to unity as well as insight. While drawing a line the child says immediately, for example:

"I draw an upright (*senkrecht*) line, one square long from up to down."

"I draw it from down to up," * and so on up to five lengths.

Or the designating words may be allowed to follow the act, for example:

"I drew an upright line (or several)," etc.

Even this little difference in perception and description (language) arouses the child's attention.

But observant action leading to consciousness is always the expression of the child's activity even when he is rebuked as thoughtless because he has turned his attention to another side of his activity than that required by us. And so we see here already how the drawing itself, even as a creative activity, and the developing, educational course of

---

* A child in a kindergarten said of his own accord, and actually with precise designation, "raised right up."

drawing evolved therefrom is necessarily conditioned in the nature and in the course of development and cultivation of the human being—child as well as man.

These simple exercises and representations, which are used first of all, and which we have just mentioned, lead also to the perception and conception, even to the recognition of general valid laws of formation and of actual life, and the all-sided power of creation increases more and more in the child the more the insight into those laws develops. But those laws reveal themselves even at this first stage (at least to the developing educator), as it were, of their own accord, with quiet observation of the thing. The educator needs only to give the correctly designating word to that which appears actually before him, and it is then possible for him to show (to the child who is drawing) these laws in the child's own action, and by that which he has represented, although this showing is yet more valuable with the fulfillment of the next requirement, viz., the connecting, comparing grouping of that which was in the preceding (that is, the first) stage represented singly, by the comparing grouping of the five vertical lines of from one to five lengths side by side.

By employing the simple rule already many times practiced by the child himself, and thus experienced by him in his own action (viz., that there can be something which is opposite to, yet like each thing), the fulfillment of the preceding requirements gives four forms, all of which have the property of being opposite to, yet like one another individually, and in the grouping of the several parts.

As now the laws of formation and life here revealing themselves (for example, that there is always an intermediate between each two things which are opposite, yet alike) are higher than those before stated, so also are the laws of development higher (that is, more general, more comprehensive), which result from the combining of the four groupings opposite (in the order), yet similar (in the length of the lines), by arranging the groups by an opposite, yet similar reference to a (common) middle. In each previous form the succeeding one is suggested. In the new combination the four groupings, which were before isolated, now show themselves as necessarily conditioned members of a higher whole.

The child's desire for signs awakens and nourishes (according to progressive laws full of life) the capacity to form a whole, to recognize the individual as a member of a whole, to find a mediating connection between opposites. In the recognition and acknowledgment of law the child feels and perceives the growth of his creative power, and soon turns the law of development (recognized, or at least anticipated in his symmetrical activity), if not to the representation of living objects, at least to the invention of more free, independent forms created from himself.

These first results and expressions of the child's power and gift of invention (by means of a definite law) are at the same time the first proofs of the creative power dwelling in the child, and indeed quite inseparable from his nature. Therefore it appears here as a requirement, which has already been quite definitely brought forward, how not merely laws of formation and cultivation, but truly comprehensive laws of life, develop from quite simple, little activities in accordance with equally simple laws of progress which are as natural as they are necessary, and more and more impressive as well as comprehensive. These laws of formation, etc., reveal themselves to the child in the simplest, actual perceptibility from his own action and representations. By the contemplation of objects, though by no means as yet by word and insight, the child is here clearly shown that his spirit is thoughtfully creative, and creatively thoughtful. At least he acts in accordance with this idea, being early impelled to such action by his own nature.

May all educators carefully consider that the observing child has passed from the total contemplation of the object as a whole to a perception of the individual parts, to the limiting lines themselves, seeing them also as single in the combination. Therefore in drawing and uniting simple lines with lines he first represents outwardly what he immediately recognizes as a definite something. In the second place, the child shows himself by means of the drawing as a being gifted with determined, firm will. In the third place, as he shows his will by definite deed, a precise something, so he also designates will and deed by definite speech and definite word. In the fourth place, he forms himself by this means into a creative being, and by these means recognizes himself as such. In the fifth place, the child brings directly into action and thought and to independent perception (though not as yet to insight) the simple, limiting, fundamental laws of form and life which exist within him.

Thus, in the sixth place, the child, as a recognizing being, rises to self-recognition, to self-consciousness. And in the seventh place, and lastly, as the completion of the circle and, in a certain sense, a return to the beginning through recognition, will and deed, self-perception, self-knowledge, and self-creation, the child forms himself into an inwardly united life-whole, into a part-whole of the all-life, into a true genuine human being. Hence it follows that the child's desire for showing and forming, his power of creating by drawing, should not exactly be freely used to produce indefinite images, but should be developed according to the laws of cultivation inherent in its nature. It is already sufficiently evident what fruits must result from such developing training, not only for the child, as before intimated, but yet more for the family, the home, the school, and for public life.

Yet the most important result, which although later proceeds with certainty from this developing means of educating the child, is that he attains to the recognition of the opposite, not as something contradictory or destructive, but as something which, conjoined with its contrast, forms an inwardly united whole. It may therefore be said that he comes to the knowledge of the opposite as being in certain respects indeed opposite, but in other respects like, and therefore actually opposite-like.

The child who, in the stage of unconscious impulse to formation, advanced in his drawing from the round to the straight, now, in the stage of consciousness, rises from the straight to the round, to the circle, which is opposite to, yet like the straight. For, although the round (the circle) is the complete opposite of the straight in regard to the law and manner of its direction as returning to its starting point, etc., yet both round and straight, in regard to their stability, to the proportion of the directions once begun, to the holding fast to the law of origination once chosen, are like one another; consequently they are from this point of view opposite to, yet like one another, but they are also alike in the fact that they are both lines which, as such, necessarily separate. But the round (the circular) line at once concludes and includes; it has its end in itself. The end is pointed out by the beginning, and thus its size is pointed out by itself. The straight line, on the contrary, running on to infinity in the direction once indicated by it, consequently, only departs from and never returns to its starting point. Thus both lines are here again opposite to one another, and therefore from this point of view also they are opposite to, yet like one another.

In the drawing and by means of it the child represents in union forms which are opposite yet alike, the ending and endless, the visible and the invisible. So, through the drawing qualities which are opposite to, yet like one another, are harmoniously developed in the child. So, above all, that which is infinite, invisible, united, existing, godlike, develops from that which is finite, visible, individual, apparent, earthly, thus corresponding to that which pertains to humanity, and so to the worth of man's nature.

Thus, in order that the child's desire for showing and drawing, consequently his impulse to formation, be fostered according to the laws of Nature and life, be developed to true power of formation, and be raised to conscious creation, the child, and thus the human being, is set in the midst of the collective life-whole and is comprehended in that life-whole. In this way is fulfilled the first condition necessary to his development according to earthly possibilities for all positions in life, viz., that he, in the completeness of his living expression of himself, be at the same time most capable, as a whole and member of humanity, of perfectly manifesting the nature of humanity; for this mode of teaching draws together the life-bond between creation, created, and Creator.

The cultivation of the child for creative (that is, independent, inventive) drawing, however small its circumference may be (for it, like a circle, is always a whole in itself), is therefore also the middle point, the starting point, and spring, as well as (according to its nature, as a condition of all development) the point to which all true, satisfying education refers. Hence the genuine kindergarten, animated by this conviction, leads to this point through each of its actions, even the smallest. Just in this cultivation of the child for creative drawing consists the nature of the kindergarten. By means of this cultivation its place of development is a garden of freshly springing humanity, and its education one of all-sided union of life, a training of the child to that which he is and which he makes known by his desire for signs—a creative being.

III.

EMPLOYMENT OF CHILDREN.—GUIDE TO PAPER FOLDING.—A FRAGMENT.

A continuously instructive employment which educates by developing, intended for children from five to seven years and over, under the intelligent co-operation of guiding adults.

Preface.

The means here presented of training the child which develop him on all sides have also the already recognized beneficent property that they, in their logical application (proceeding from the simple, and constantly progressing toward the more complex and manifold), offer an agreeably refreshing and strengthening recreation for the children, and not less for the power of mind and spirit than for the bodily activity of the child-tending adult. These plays possess the property of satisfying alike the playing child and the loving, guiding adult (mother, father, older brother or sister, educator), and of leading the adult himself to an intelligent enjoyment of them, and this is just what makes them so appropriate for the genuine family room, for intimately united family life, and for fostering such life. This is the property which causes these childish employments to act so beneficially on the uneducated child, as well as on the educating adult, uniting both to form a whole, full of life, for the practice and representation of the true, the good, and the beautiful, as they develop spirit, mind, power of doing, thought, feeling, and action in harmony and accord of life, which is such a great, imperative need for all our relations in life, and for the all-sided demands of life for intelligent, æsthetic, religious, practical training.

A.

A guiding thread through the means of employment in general as well as a superintending guidance in particular.

1. As was demonstrated in the Mother-Plays and Nursery Songs and elsewhere, the means of developing the child are deeply conditioned in his nature and his course of training, proceed directly from the bodily activity and the spiritual influence of man as a whole, single in itself, and exert a direct influence on man. So the means of employing the child, presented especially in the so-called play-gifts, etc., are, as can be demonstrated, as deeply grounded in the nature of the child as in the properties of the things (as has also been re-

peatedly said here), and proceed from a connecting medium, from the corporeal, the material, as such, from its simplest representative (the ball) to the sphere, the cube, and the cylinder as an undivided whole; since the children consider and especially treat these objects as such, according to their various and opposite as well as connecting properties and relations in rest and movement. The training of the limbs and senses is here the connection between development and employment, the single and the undivided, the spiritual and the corporeal.

2. The advance from the undivided is, according to the ruling law of opposites, to the divided, and in the divided, first from the once divided on all sides to the several times divided, from the cubical to the brick formation, and from the straight to the oblique division. These are the well-known play-gifts from the third to the sixth inclusive.

3. From the bricks (of the sixth gift) the means of play goes on to the tablets, and, in the tablets, first to the right-angled or four-cornered, and then to the oblique or three-cornered. The comprehension of their different positions and relations to one another conditions the difference in kinds, their use, and their different effect on the child.

They form the special and intimately connected series of the laying plays which comprise a many-sided cultivation, and hence are especially adapted for developing education as well as agreeably instructive.

4. Through the brick formation and that of the tablets the corporeal division goes on to the sticks; first of all, to the sticks of the length of two cubes, or one brick. The use of these sticks consists:

a. In different ways of laying the forms of life, beauty, and knowledge which are already abundantly familiar. With the latter we can again, if we wish, distinguish the figure and number forms, and also the writing and reading forms, etc.

b. In the firm joining of the sticks by a connecting body (peas or cork) to form inclosed figures or surfaces.

For movable connection of the single sticks there appear: 5. In this place, slats of the length of eight cubes or four bricks, and of the width at most of a half cube. The employment with these is the so-called interlacing. The preceding employment is called peas work.

6. The connection of the sticks by the interlacing, appearing as an undivided but movable whole, leads to the employment with the jointed slat, by the management of which one form is developed from another, and the more complex is brought back to the simpler form.

7. The interlacing and the use of the jointed

slat lead to the employment with undivided strips of paper folded together lengthwise, or to intertwining. The joining and twisting of these proceed from simple forms and conditions, and lead then to the combination of single colored strips of paper in a turning fashion (braiding). This leads to weaving, which opens the passage to the following.

8. Through all which stands before us, the division of the whole and its separation into parts leads back again to the whole in perceptions of solids and surfaces, but both can be seen through, and thus permit a view into the interior. From this division and recombination of the divided material the advance is to

9. Changing the material in different forms, but in unaltered quantity, to wholes, which actually remain constantly the same in themselves.

a. If the whole, when it is to be changed in this way, is massive, the mass must be soft, capable of being pressed and moved out of place (thus in a certain sense impressible), therefore changeable in form, but abiding in respect to quantity. This is the modeling in its first simple forms, proceeding from the sphere or from the cube.

b. If, on the other hand, one proceeds from a defined surface which always remains the same, the result is the folding, which is to be especially treated and brought forward in what follows.

c. If one proceeds from a thread there results a childish and especially girlish play, the so-called thread game. Here are brought forth the outlines of surface and solid forms of different kinds by the use of a thread (the two ends of which are joined, and the length of which remains the same through all changes of form), with the help of the two hands of a coadjutor and the use of one's own hands.

10. In accordance with the course of constant development, the childish employment now following is of course the connection of the two preceding ones. It is therefore the union of the separate with the abiding, or, with the solid mass, the cutting off or carving; with surface material (paper), the cutting out—cutting, so called.

The cutting is therefore the connection, the folding of an abiding material with the parts and forms, so that an abiding material indeed remains, but develops and shows definite forms in the relations of its separations; although also, on the other hand, separated parts again originate which may be again combined according to general and necessary laws of combination.

But from the management of this fundamental activity there proceeds again a twofold course: First, one may regard the forms which are originated by the cutting in respect to their kind. They may then be either merely representations of the beautiful, of an idea; or representations of facts and thoughts; or representations of the forms and objects of the surroundings, of Nature and of human life—even the representation of man himself—here, for example, the furniture of a house, an inn, a manufactory, dolls, human forms; there, natural objects, animals, plants, etc.

Or, second, that which is cut out, cut off and separated, remains in connection with the whole, and is itself grouped together to form a whole, inclosing space. This gives the hollow, and thus in its application to human life proceeded the room, the house, the building; and thus again the introduction and return to the purely human, to the home and the family life, to the family room, from which indeed the whole proceeds. All this is presented to the child in subjective form, and can be understood by the child from his own experience.

Thus, then, a review of the whole employment and occupation of children shows that they are to be, for the child, a mirrored image of the whole life of Nature and of man, and that at the stage of childhood they lead to true estimation and comprehension of life; indeed, at every stage of the development and the progress of childhood they lead to comprehension of its inner significance.

From the sequence of the employment and occupation of children here intimated we now bring into prominence but one—the paper folding.

It proceeds from a surface of a precise and given form.

This simplest form (proceeding from its innermost foundations and law of development) is the square, or the form defined by four equal sides, four equal angles, and four equal corners. Although the triangle, consisting of three equal sides, three equal angles, and three equal corners, appears, on account of the number, to be the simpler and first, yet it is, on account of its nature (as is afterward shown), the one which is led up to—the later.

B.

Paper folding as a means of employment proceeding from the square surface or square form.

The square may be formed from any firm paper surface, whatever the outlines of the latter may be. The given continuous surface is creased by a fold into two parts as nearly alike as possible. Then the surface is again doubled together, dividing into two equal parts the line caused by the folding, so that a right angle results, closed on one side, open on the other. Then one half of the open side is bent so that its boundary line coincides with the boundary line of the closed side, this, with the new fold, forming half a right angle. We then make an incision at any point of the sides which have been laid together and made even all the way from the apex of the half right angle, therefore on the side where the half of the divided side coincides with the closed one. We then let the half fall back into its original position and join the two incisions by a straight line in the direction of which the superfluous paper is cut off. Now if the triangle resulting from this cut be unfolded the desired square appears.

The remarkable thing in this is, that the most symmetrical and simplest form, the square, results from the unshaped surface by means of three creases and three cuts. This phenomenon demands the strictest consideration on several sides.

By the foregoing work and its results should be demonstrated that the formed, and, in this especial case, the square, proceeds from the unformed by regular division.

But it would be not only unadvisable, but ridiculous, to proceed always in this way to form a square. A proper, progressive cultivation always takes into account what already is. If not square forms, we find enough rectangular forms in our surroundings. These are shown us by our present machine-cut papers, and often with great exactness. This is a beautiful and convenient agreement of that which already is with that which we want, and we must thankfully consider it, and make use of it to promote the whole. Thus if it is desired to carry on the folding with additional paper it is advisable to take for the purpose rectangular machine-cut paper. Ordinarily, for instance, in the work of the public schools, paper already used in writing or drawing books can be taken for the folding.

We here proceed with the machine-cut paper as material for folding, or, in other words, we make use of such paper. The use of already written paper, or of leaves from writing and drawing books, proceeds on the same plan.

We see, first of all, how from one sheet of machine-cut paper a number of suitable squares can be made, as preparatory material for folding.

In order to understand what follows, and to be actually instructed by it, it is necessary to take a piece of machine paper in hand and follow out the directions step by step. If one has anything clearly before him, it is easy to impart it to others.

1. I lay the whole sheet before me opened as a horizontal rectangle, crease it by a fold in the middle of its length so that it is divided into two equal parts or oblongs (one long side of which is closed, the other open); the two shorter sides are divided. The paper thus folded lies before me in the position of a horizontal oblong, the closed side turned toward me, the open side from me.

2. I take then the upper half leaf of the divided right-hand side and bend the corner so that the

short side of the oblong shall coincide with the closed side, and so that the right angle which the shorter side forms with the closed longer side is divided into two half right angles.

3. I turn the whole over, so that what was before underneath shall be above, and do the same with the right-hand corner of the top leaf now lying before me.

4. I do the same with the two left-hand sides. The result is a parallel-sided, alike-sloped quadrangle (boat-shaped trapezoid).

5. I unfold the whole in the crease which formed the two original oblongs, through which a hexagon appears, two of whose sides are single (in thickness), the four others closed by the four bent-in triangles. The bent-in triangles lie two on each side, and so that the bent-in points coincide in a straight line, which is at the same time the base of the large triangle formed by the two small bent-in equal triangles.

6. In the direction of these two bases I now bend each of the two large triangles back, press the two folds close with my finger, and cut off the triangles in the direction of these folds.

7. Each of these two cut-off large triangles, each of which again consists of two doubled small triangles, I bend together in the already creased fold, so that the two doubled small triangles lie one on the other, and cut them apart in this fold.

So I obtain four doubled small triangles, and, when I have unfolded each of them, four squares.

8. There remains yet of the sheet one rectangle lying in the middle containing equal quantities of the two half sheets of paper, which will be used at a later period (for the intertwining). It is essential to consider the position of the knife in cutting; the edge of the knife must be on a line with the paper, and the cutting must be done very firmly.

The squares resulting from the foregoing, and by the repetition of this with several sheets, give now the foundation exercises for the folding, and for the forms and representations resulting. These forms are different indeed, but develop from one another in simple symmetry, first, however, making several additional symmetrical folds and creases, by which a form results which is the fundamental for all following forms. It is well for the children to make a number of squares themselves before carrying out the folding, and to represent with them a number of fundamental forms. All this aims at and unites perfection, by which the fresh and glad continuance of an employment is promoted. In addition be it said that it is better to call the several-times-mentioned figure of the surface a square, rather than a quadrate (*Geviert*), because four equal sides, four equal corners, and four equal angles are found in it.

We later call another figure a threefold form because it is formed of and bounded by three equal sides, three equal angles, and three equal corners. We name each similar figure in the same manner, by which means an equally progressive nomenclature for all equal-angled and equal-sided figures is found and given, instead of the designation consisting of many words hitherto used—"regular pentagon," etc.

From the stock of squares now obtained result the fundamental forms in the following regular way:

9. In order constantly to develop the artistic from that which lies before you, the square is taken between the two hands so that the forefingers lie within the already creased diagonal fold, but the thumbs and the middle fingers of both hands must be outside the fold. Now draw out the two forefingers gradually from the square and crease it again in the already existing cross fold into two equal parts, saying: "I divide the square by a diagonal line into two equal parts, into two equal and similar right-angled, isosceles triangles."

REMARK.—The fundamental law of all advance, development, and cultivation (thus, in general, of all education) is to proceed from any given thing to the pure opposite within this given thing. So here also.

9

Opposite to the oblique lies the right line, expressed in the square as across.

10. I lay the square between my fingers, as before indicated, so that they now lie in the middle of the square, of which two opposite sides are laid together (as were before the two opposite corners), and say, proceeding in the same manner as before, while I lay side closely to side and corner to corner: "I divide my square by a cross line into two equal parts or halves, two equal long rectangles" (evident by superposition).

I now unfold the whole again, lay my forefingers, as in the beginning, first in the diagonal line (a), and then in the same manner in the cross line (b), saying with "a" "one half is," and with "b" "equal to one half" (and going back to the first position, "a"), "consequently the triangle" (now going back to position "b") "is equal to the rectangle."

REMARK.

1. By this word and speech connected with deed by the adult, the child will at first only be required to do and to hear, because the organs of speech at this stage of the child's development are as yet too untrained for the repetition of long sayings, and the mind is as yet too unpracticed for comparing comprehension. This practice will be obtained in the way just mentioned. It is worthy of consideration that the child appropriates the

words more easily by frequent hearing than by frequent repetition, for hearing impresses the mind more than repetition. Therefore everything must be clearly and precisely expressed by the adult always with reference to the perception.

2. With these exercises various perceptions of the relations of form and size and facts for the true pleasure and elevation of the life of the children may be connected, in proportion to the development of the children; for example:

a. A square will be divided by a diagonal line, more exactly indicated as a corner diagonal, into two equal parts, two equal right-angled, isosceles triangles, which are opposite to, but like, each other (that is, while their two times two equal acute angles lie at two opposite corners, the two right angles and base lie likewise opposite to and equal to one another).

b. From this follows the perception of the fact that a triangle is the half of a square when it has an equal base and altitude. A further perception is this:

c. A square is divided by a cross line which goes through the middle of the square parallel to two of the sides, into two equal parts, two equal rectangles. Further,

d. This cross line divides at the same time each of the two sides at its ends into two equal parts. By the diagonal line the perception is retained that,

e. As the diagonal line divides the square into two equal parts, it also divides each of the right angles from which it proceeds into two equal parts, so that each of these is equal to half a right angle, and both together again form a right angle. From this follows further the perception of the fact that,

f. The sum of the angles of a right-angled, isosceles triangle amounts to two right angles; and the angles at the base (that is, at the diagonal line) are, taken together, just as large as the angle which lies opposite to the diagonal.

By these employments, which are plays to the child, and which can be made very attractive to him by the quick change of form and the quick indication of that change by accompanying words, the child also perceives that the expression right angle has a double signification: first that of form, then that of size, in which lies its connecting—that is, instructive—property and importance.

From this follows the seventh easily demonstrable perception, when the square is divided into the two triangles,

g. That in a right-angled, isosceles triangle the base is always larger than each of the two equal sides or legs of the triangle.

REMARK.—All these facts, viewed here singly and as separate phenomena, receive their clearer and more general perceptions in the progressive development of the whole from stage to stage, by their frequent repetition in the most various relations. They impress themselves so deeply and easily on the mind as perceptions of fact and real truths that by the retention of them the abstract words do not burden the memory. The retention of the so-called abstract—that is, purely intellectual—facts is connected with the pleasure and joy of activity, of creation. The child finds them, during the pleasure and joy, in himself and through himself, and so easily receives them as his own self-won property; for, briefly speaking, they would not be impressed upon him, but would develop in him by his own activity. We now return to the perceptions of knowledge.

h. A right-angled, isosceles triangle is equal to a rectangle which has an equal base and half the altitude.

The perception of this truth proceeds from the two kinds of division of the square by a diagonal line and a cross line, and by folding the paper together in both ways for the purpose of comparison. (See above, No. 10.) "One half is equal to one half."

But this fact can also be expressed as a new one in the opposite way—that is, by proceeding from the observation of the rectangle, as before from that of the triangle.

i. A rectangle is as large as a right-angled, isosceles triangle which has an equal base and twice the altitude. This can be made perceptible by applying the diagonal line of the triangle, cut off from the rectangle, to the second half of the divided side line, and thus in twofold, similar directions joining a right and an oblique line. Further perceptions are received when I spread out the square before me and consider the relations of the two creases inside the square.

k. The diagonal line and the cross line form, at the point where they intersect, two times two equal angles, which, at the point of intersection, are opposite to and equal to each other, and which are called vertical angles, from which is received the perception of the fact that vertical angles are equal.

If now we take into consideration the two sides of the square, which are parallel with the cross line, in connection with the diagonal, we obtain the following new perceptions:

1. The perception of inner opposite angles, here twice formed from one of the obtuse vertical angles and the acute angle lying opposite to it which is on the same half diagonal, and is formed from this and the side of the square which is parallel to the cross line.

2. The perception of inner, alternate angles twice two times formed: once from the diagonal line and the two horizontal side lines of the square, and lying on the alternate sides of the diagonal;

then formed from the two perpendicular sides of the square and the diagonal and lying on the alternate sides of the latter. Then again two times two, formed each of a half diagonal with each of the two half cross lines, and with the two sides parallel to the cross line, and lying on alternate sides of each of the half diagonals. The inner alternate angles here indicated are all sharp, but the right angles can also be perceived. But this intimation must suffice. Since now each pair of such inner alternate angles are always equal to one another, we obtain the perception of the fact that—

l. Inner alternate angles of parallel lines are always equal to one another.

Other positions of the angles can be perceived; indeed the children seek them for their own pleasure after their eye and sense of form are developed.

We will now go on with the exposition. We last obtained a rectangle, one long side of which was closed and the three other sides were open. We now bend this together so that one half of the closed side lies on the other, and the two short sides of the rectangle come together. This gives a square with one wholly closed side, and one half closed; the two other sides are open, and in four parts. I open the square again to a rectangle, and obtain the following perceptions:

m. A short cross line, which goes through the

middle of the rectangle, divides it into two equal parts; each of these two parts is a square.

n. Inner opposite angles of parallel lines can both be in respect to form two right angles (perception of fact), and are together two right angles in respect to size (statement of fact reached by deduction).

The two long sides of the rectangle form, with the separating cross line, two times two right angles; or conversely, the separating cross line forms, with the two long sides of the rectangle, two times two right angles. These angles lead to the name —adjacent angles of a line—and this leads to the perception that in the case in question two adjacent angles of a line have the form, as well as the size, of a right angle.

We now open the rectangle again to a square and thus come to the perception that—

o. Two lines which cross a square and intersect in the middle, and each one of which is parallel to two sides of the square, divide the square into four equal squares.

This perception comprises the fact that—

p. Each component square is one fourth of the principal square, and so each small square is like the others in form, as well as in extent. The opposite statement, that each of the small squares is like the principal square in form, but different in extent, gives rise to the following:

q. Like form does not condition like size, or the size can be different with the same form. This fact can be again brought out by the comparison of the four small triangles with two large ones.

This perception of the two cross lines connected with that of the diagonal line passing through their point of intersection—by which, according to form, there originate two right and four acute angles, or if we bring only one cross line into combination with the diagonal, two obtuse and two acute angles—leads to the perception that—

r. All the angles round a point taken together equal four right angles.

s. Adjacent angles—that is, such as come together on one side—are, according to size, always equal to two right angles, for they take up only half the angular space around a point.

t. If the adjacent angles are oblique, the obtuse angle is as much larger than a right angle as the acute angle is smaller.

REMARK.—The child must be repeatedly made to observe that a right angle is spoken of in two different ways. When, for instance, I say, " the angles may be of two kinds, right or oblique," I speak of the right angle according to form. But when I say, " a right angle takes up a different space from the oblique," I refer to the size of the angle. The name, right angle, thus refers (1) to the form, and at the same time (2) to the size of the angle. In this double nature and significance of the word *right* is founded the fact that the right angle is in itself the measure of the angles, for the right angle, on account of its double nature, forms the connection between the obtuse and the acute angles.

u. As was formerly said with respect to the adjacent angles perceived in the folding leaf, the obtuse angle is as much larger than the right angle as the acute angle is smaller, so we here perceive that the two obtuse angles are as much larger as the two acute angles are smaller than two right angles.

In this way the eye of the child is developed to the perception and conception of a number of relations and facts in proportion to the child's skill and capacity for inward and outward perception. The eye is spontaneously and the mind voluntarily developed for such relations and facts by the frequently repeated representation of the thing, since word and deed, perception and designation, and thus thinking, doing, and noticing, are always intimately united.

We now go back again to the occupation of folding. The last fold was that by which we obtained seemingly one square, but actually, after unfolding, four squares by the division of the long rectangle. We now again connect with the first experiment where a square was made from the rectangle by folding the latter together in the shorter cross line. Now I fold the lower closed side to the left side, and say, "I divide the square by a diagonal line into two right-angled, isosceles triangles."

Now, turning the square thus divided downward, and the opposite one upward, I do again as before and say: "I divide the square into two equal parts, and, indeed, into two right-angled, isosceles triangles."

I now open the rectangle, by which means a larger right-angled, isosceles triangle lies before me, as well as the rectangle. From the simple perception of facts proceeds the double truth that the triangle is the half of the rectangle; and, looking at the two in the opposite order, that the rectangle is twice the size of the triangle, which perception leads, at this stage, to the general saying that—

v. A right-angled triangle is half of a rectangle which has an equal base and altitude; or, reversing the statement—

w. A rectangle is twice the size of a right-angled triangle which has an equal base and altitude. The fact, as it lies before the child, expresses this truth so clearly that the word only makes an audible truth of a visible one, or an advance from the outward perception to the inward recognition and comprehension of the truth.

The performance of the thing proves the truth and importance of what has just been said for the fostering of childhood by developing and educating. A further perception (twice shown, first in the separation, and the second time by the crease or fold, serving, as it were, to confirm one another) is that—

x. A perpendicular from the right angle of a right-angled, isosceles triangle to the opposite base, divides the angle, the base, and the whole triangle into two equal parts, the two halves of the triangle being right-angled, isosceles triangles.

From this saying there can be again derived a large number of sayings which are the result of perceptions, although they are derived from or connected with the preceding ones.

When the surfaces of the divided squares lie side by side, the rectangle thus formed is again divided into two squares for the children, with the familiar words, "I divide," etc. Each of the two squares is likewise divided by a diagonal line, using the same words as before. By this division is obtained, apparently, a single right-angled, isosceles triangle which really consists of several such triangles. By unfolding these triangles we have before our eyes, apparently, a square, but consisting of two squares lying one upon or one behind the other, one of which is divided by two actually separating cross lines into four equal parts (four right-angled, isosceles triangles), and the other square, now brought forward from behind, is likewise divided into four equal parts, but, by two cross folds, into four equal squares. This experience now gives the proposition—

y. That two diagonal lines divide each square into four equal parts, four right-angled triangles, and

z. That two cross lines [through the centre] parallel with the edges likewise divide each square into four equal parts, or into four squares.

This leads to the concluding statement:

aa. A fourth is equal to a fourth; consequently each of the squares is equal to one of the triangles, and each of the triangles is equal to one of the squares.

If now the paper be opened, we have again the first principal square. Inside of it is another square, in an opposite position to the principal one, and therefore called the opposite square. Its corners or angles lie in the same direction as the edges or lines, and its sides in the same direction as the corners and angles, of the principal square. So it seems to us the name of the opposite square is fully justified. The comparison of this with the principal square gives rise to the following truth of perception:

1. The opposite square is half of the principal one, and, reversing the statement, the principal

square is twice the size of the opposite one. From this we derive the statement that surfaces of like form can have quite different extent, and, conversely, surfaces of different size can have like form. A further statement is that—

The four triangles outside of the opposite square are, taken together, just as large as *it* is, and each of the triangles is a quarter of the opposite square. The opposite square is four times the size of each triangle which touches one of its sides. Now comes the new perception that—

2. Two cross lines, each of which is parallel to two sides of the principal square, and which go through the middle of it, divide it into four equal parts—four equal squares. Each of the triangles outside the opposite square is half of such a square; thus half of the quarter of a whole. The half of a quarter of a whole is an eighth,* consequently each of the triangles outside the opposite square is an eighth of the principal one, and so a fourth of one of the triangles we first obtained by diagonally halving the square.

All these facts lie directly before the child in his play-employments, and it is merely necessary to give the word for the perception; not that the child may retain the word, but that through

---

* The perception of this fact has been prepared for by the earlier play employments, especially with the third and fourth play-gifts.

the word the perception may become an abiding one.

In like manner, several more propositions, which are the result of perception, are derived from all the preceding statements, but are left for individual discovery.

We now go back to the double square (that is, the one in which the four outside triangles are bent back upon the opposite square, with their right angles touching one another in the middle of the square). I take it in my hand so that the divided surface lies uppermost, and the two forefingers rest in the cross line; following this line, I begin, as before, the division of the square into two equal oblongs by a cross line, repeating, as usual (in order to bring the repeated phenomena to the child's perception, and to put them into words), "I divide the square by a cross line into two equal oblongs."

The so-formed oblong, placed in a vertical position, is now divided as before by a cross line into two equal squares, so that four doubled squares lie one on another, and so that in one corner all four are separate, and in the diagonally opposite corner all four united. I now bend the wholly separated right corner back to the closed corner, thus dividing the square as before by a diagonal line into two right-angled, isosceles triangles. In the same way I treat the opposite square. Now I bend the two triangles thus obtained so that they cover one another, and the two remaining squares are outside. Now I divide these in the same way, and open the whole, and there appears a threefold square which on the side of the diagonal line is divided into equal right-angled, isosceles triangles, and on the opposite side by two cross lines' into four equal squares. This repeats the earlier recognized perception, but with much greater clearness.

3. When a square is divided by two diagonal lines into four equal triangles, and by two cross lines into four equal squares, each of these squares is equal to one of the triangles; and, reversing the statement, each of the triangles is equal to one of the squares, so that each of the triangles, and each of the squares, is equal to one quarter of the whole square. This is one of the finest, clearest, and most cultivating perceptions for the child. This gives further the correct perception—

4. That the two cross lines and the two diagonal lines intersect each other in the middle of each, and so bring the middle of each and the middle of the square manifoldly to the child's perception and knowledge, and, as it were, reveal it. For, from this point on, the middle now appears manifoldly important in reference to the outside; and in the reference of the outside (the single) to it, the unity.

With these four squares so obtained (by the comparison of which with the triangles, all the earlier perceptions are repeated with increasing clearness), the first fundamental form is given from which the first principal forms necessarily develop.

IV.

STICK-LAYING.—A FRAGMENT.

As has been already many times intimated and presented, to a greater or less extent, this whole of plays and occupations has by no means originated with me arbitrarily and as an artificial mechanism outwardly thrown together; but it is actually, in all logical consequence, called forth, and, as it were, given by the stage of cultivation of life now begun, and universally verifying itself and aspiring to become common property. This whole is also called forth by the idea of education and training of the human being which is developing itself to greater and greater completeness and applicability. One may say, indeed, that this whole has grown forth with such necessity as, in the spring, the seed germinates, the bud swells, and finally the tree blossoms and bears fruit. Even the personality, through and by which this whole appears, has not been able to interfere with its pure, logical consequence, or with the fundamental conditions of its perfection. Thus the whole is an equally necessary, fresh, and healthy growth of human development. Although the statement is now revealed in many ways and confirmed by experience without the possibility of doubt, it may be somewhat anticipatory to say that the wonderful efficiency of the whole of the school stage, as well as that of childhood, proceeds from its harmonious growth.

Consequently we enter into the subject so as to let it unfold itself according to its own inner law before our eyes. According to the measure of the facts so resulting, we can then demonstrate the more extended results of and demands for the developing educational training as the fruit of the whole in the proper place for further consideration. In so doing we view such training as the fundamental effort of the present day and the characteristic of the present state of cultivation of humanity.

As is the case with each of the means of play, or each plaything which makes its appearance in the whole of plays and employments, the sticks also, the stick play, and thus the actual stick-laying, are by no means arbitrarily or accidentally introduced into this whole; but the stick-laying steps forth in the whole of plays and employments, like each of the other plays, with inner necessity, at its precise place, in its peculiar way, and (remarkable!) at once in an age and at a time in the life of the child when he has attained on every side to the power and capacity, not only to play with it in the usual sense of the phrase, but actually to employ himself with it thoughtfully, and therefore formatively. And thus, after the beginning of the fifth and in the fifth year of the child's life, the stick-play grows up in increasing completeness, even as the child himself grows older.

From this fact can be clearly deduced the statement that if it be desired that the stick-laying have a healthy, invigorating effect on the life of the child, it should not begin before the child has obtained the requisite preparation for it by means of employing himself with the plays which precede it. The capacity for conception, remembrance, abstraction, and creative representation must have already attained a certain power and cultivation in the mind of the child, or else there will come forth merely insignificant and immature results for this stage of the child's life, and his employment with this play at this premature stage will bring him more injury than gain. The child soon forms the opinion that because he, with his as yet too undeveloped weak power, can accomplish nothing with this means of play, nothing can be represented by it, and he therefore treats it with too much indifference. But, on the other hand, it should be remarked that the sticks are a means of play, and a material which can be easily managed by a relatively weak power.

In the series of the whole of divided plays and employments, the sticks proceed from the tablets, which, being split in the direction of their length, can be divided into sticks, and, as it were, fall apart into such sticks. For this purpose it is best to take tablets made of pine or fir wood.

Peculiarly developed, vigorous, adroit, and careful children of this age may be allowed to split the tablets themselves, and they like to do it if given corresponding tablets for purposes of comparison. The effect is, first, the children in a certain sense create their own means of play and employment, and, second, from their own action discover how one means of play develops from another. Thus the connection of things is made plain to the children by and in the continued development, and, essentially, by means of their own activity. These are the most important and formative properties of this means of play, and will later manifest themselves more definitely with other employments of children.

On account of its importance let us now bring ourselves to a clear perception of such a development of the later from the earlier, of the newer from the older, of the last from that which immediately precedes it, in the manifest development of the sticks in our play-whole, so that we may be in a position to bring to comprehension and insight such observation of the development of our playing child by this means of play and employment.

The sticks, as we have just said, result from the tablets.

But the tablets are, in like manner, a progressive development of the oblong prisms of the fourth play-gift with the three different relations of size, according to length, breadth, and thickness; the length being two cubes, the breadth one cube, and the thickness one half cube.

The oblong prisms in the just-mentioned relations of size are, however, again developed, according to the same law, from the four rectangular, four-sided, and equal-sided columns into which, in the course of expounding the third play-gift, the cube separates, so that thus the oblong prisms proceed from the cube through the equal-sided, four-sided columns.

But the cube is a quite necessary development from the sphere, as the sphere is, for the more developed child, the firmer, and, for that reason, the more movable, and thus, as it were, the more perfect ball, which fact has been already brought out by the development of the second play-gift.

But in the ball, and yet more definitely in the more rigid sphere, can be perceived (as opposite to the external round surface) the straight lines, first of all in the three principal directions which intersect each other at right angles.

These three invisible but yet definitely perceptible directions in the ball—which is, as it were, the germ of all, the developing means of play and employment—and yet more in the sphere, appear in the cube as three times four straight edges; in the rectangular, equal-sided, four-sided columns as four times four; and in the eight oblong prisms as eight times four straight edges; and finally these become, by splitting the tablets into sticks, a multitude of sticks.

And thus we have developed the stick backward from the sticks to the sphere and ball, and from them forward to the sticks. We have thus treated the sticks as a means of cultivation in that kind of developing education which requires thoughtful observation and judicious accomplishment. This is important in order to show how the sticks in their first appearance and actuality, as straight directions and lines, are already given (drawn into a smaller compass) in the sphere and ball, and have in them, as it were, their germinating point and root, their first origin.

As then this whole of plays and employments, being founded on fact, actually develops in the purest logical sequence, and necessarily from the sphere and ball, the latter may be considered as the germ of a tree, and all the means of play and employment may be symbolically regarded as part of the tree developed to blossoms and again ripened to fruit and seed, a figure of speech already used in other articles on this subject.

Or the ball or sphere may be considered as a flower bud which develops from itself in the blossom a great number of stamens and pistils.

This symbolic representation of the unfolding of the whole in the figure of a tree unfolding itself in like manner, and in the blossoming and blossomed flower buds of such a tree, is indeed to show and demonstrate for the child the constant (consequently full of life) development of each individual part of this play-whole (and, above all, here of the sticks), as well as the nature of the employment which enters the whole at the right place. This development of each individual part of this play-whole lies in the nature of the thing; for it is important even for the child, but yet more for the total development of the life of the complete human being, that the child should be early led, even through his yet sportive life, into these inner linkings of life in a chain of which he is himself a link, in a manner corresponding to his intellectual power and proportioned to his bodily strength. Such guidance is the chief object of the training which educates by developing, and of this whole of plays and employments which is, as it were, composed of organic parts.

What is now the stage of cultivation in the child which the use of the sticks presupposes?

It is already a considerably developed one.

First, perfect use of the limbs and senses, especially cultivation of the sense of sight and the use of the hands and fingers, for which the Mother-Play and Nursery Songs, as a family book, gives the most appropriate guidance to the most versatile and earliest stage of the child's development.

Second, a clear perception of the round and straight; with the straight, a clear perception of the right and oblique; and with the right again, of the vertical and horizontal; with the oblique, of the right and left diagonal. The plays with the ball and sphere already give these perceptions (and have manifoldly given them to the child), which become more and more cleared and confirmed by the subsequent means of play.*

But the child has also gained the perception of other positions and directions with respect to one another, such as are parallel and such as are not parallel lines, similar slope and similar direction of lines, just as a pure perception, indicated by simple words without explanation. The ball and sphere plays, with the body and movement plays result-

---

* With all that is expressed in what next follows one should, in order to attain to a clear and correct understanding of it, have the object in question (the play-gift) actually before him when possible; or should at least try to recall it as clearly as possible to his remembrance in order to rise from the outward perception to the inner, and from the outward grouping to the comprehension of the inner intellectual coherence. Such is also the case with what precedes, beginning at the first page of the stick-laying.

ing from them, manifoldly develop these perceptions; but yet more the play with the cube of the second play-gift, through which came to the child the clear perception and conception of angles and corners, edges and surfaces, sides or planes, as connected and unconnected lines.

But these perceptions and conceptions were also shown with peculiar clearness and on all sides in beautiful arrangement and grouping by the third play-gift (the cube, once divided on all sides), which is therefore called the delight of children.

It is especially the right, clear conception, the adroit and thoughtful management of this third gift, which is presupposed by the play with the sticks, and especially by the stick-laying, a play rich in results and therefore educative in many ways; for by these perceptions of the mass the child trains itself by degrees to the perception of the outlines which are, as it were, drawn out from the mass.

There is but one perception now lacking to the child's complete and all-sided developing and representing use of the sticks. This is the perception not only of different lengths, but of proportionately different lengths. This perception has already been brought before the child many times by the third play-gift, but changeably, not abidingly, and, as it were, fixedly. But this is supplied by the fourth play-gift, the eight oblong prisms.

If we add to this the laying with the tablets, and let these, as above mentioned, unfold or fall apart into sticks, we have come with our child to the stage of development of his activity and attraction to employment when he can practice stick-laying with developing results, and can, in and by means of it, reveal himself in yet greater compass than heretofore as a creating being, developing and forming from himself.

And now what have we gained by this presentation of the stick-play, and especially of the stick-laying, in reference to the whole of the means of play and employment, as well as in reference to each single play-gift in respect to its use as well as its spirit?

First, we have recognized that this whole of play and employment is not merely outwardly drawn together in respect to its single plays, but is developed in all its parts according to inner and necessary laws, and is consequently an organic whole, full of life.

Second, that the profitable use of each following play-gift in a certain compass presupposes the knowledge and use of the preceding ones, but that after and with each of the following play-gifts the preceding ones can be used with yet richer results. In the same way in the world surrounding the child and man, the most various kinds of objects are present at the same time for use as well as for consideration, and can be actually used at the same time by the man, and even by the child in proportion to the stage of cultivation he has attained.

Third, the bringing out of stick-laying according to its spirit and use teaches us that the means of play and employment as a whole, as well as each single play-gift, introduces the child symbolically into life according to all points of view, relations, and directions, and that it is for the child a truly developing, educating means of cultivation. This will be shown more and more clearly in the course of demonstration.

Fourth, here again is presented as worthy of consideration what has already been shown us by the study of our modes of play and employment, namely, that the development and representation of the manifold presupposes the knowledge of the single; but, again, the knowledge of the single as a member of a higher and greater whole demands and requires the knowledge of this whole as its unity, without which knowledge the single can not show its nature again in manifoldness and perfection—thus three in one.

Fifth, if we now, according to the foregoing, consider the stick as a member of the whole of plays and employments, we find, according to the fundamental law of this whole, that the stick includes in itself all the essential properties which the ball contains in itself as a whole and as a member of the surrounding world of objects. Hence it has the property of filling space, that of having boundaries, material or contents, coherence, gravity, extension, but here pre-eminently in the direction of length, though in the ball, sphere, cylinder, etc., there is extension on all sides. Hence the stick has also form, size, number, color, even sound and elasticity.

But further and essentially in and by means of the visible (for example, by the two visible terminal points) there appear in the stick the invisible line of direction and the middle point likewise invisible, but sharply defined and consequently separating, and at the same time in a remarkable manner uniting. The stick, therefore, like the cube, sphere, and ball, unites the highest and most general laws of earth, Nature, and the whole world, which connect and unite in spirit. Such is the law of unity and individuality, and that of opposites and their connection.

From the middle toward each of the two end points new middles are again postulated as points again uniting middle and ends, thus as new points of connection, and so on, fixed always between each two newly resulting points, thus bringing out the phenomenon and law of continuity.

In like manner the sticks show and unite the visible and invisible. And so we see in them not only the essential properties of the cube, cylinder, sphere, ball, etc., but also the most essential proper-tics of all the objects which surround the developing child. Therefore the stick is for the little one a means of introduction (through the connection of the visible, invisible, and invisibly visible) to his own life and to the surrounding world.

Having hitherto looked at the stick by itself, we can now view it in its relations to the objects around it, and thus, first of all, to each plane or surface, even to the surface of the earth itself. The stick can either stand in upright direction toward the earth, and in this direction sink, as it were [or gravitate], directly (right) toward the middle of the earth—that is, stand perpendicular (*senkrecht*)— or it may incline toward the surface of the earth with all its points at the same time and quite equally, and consequently lie horizontally, or it may be in a position in respect to the earth, which, as it were, connects both relations—that is, in an oblique position. Now all these properties which slumbered, hidden in the stick, as it were, for the child, and others which the stick-play brings by degrees to the view of the children, are what gives the stick such an inexpressible charm and such an enchaining power of attraction for the child as an object and means of play; and, indeed, gives the child, as a creative being, the premonition that the stick affords a suitable material for his impulse to creation.

We now go back finally to the stick-play and

the stick-laying. The stick is, for the child, either a middle line of the sphere made visible, or a cut-off edge of the cube, or one of the sticks which are the result of the splitting of the tablet, or a straight line. But in a wider point of view the stick is for the child the representative of all things that are straight. And the play is connected with this simple connection of ideas, this simple mode of comprehension of the child. We see whither this quite simple way of comprehension and perception leads our child in his symmetrical development as a creative and recognizing and, we may add, as a feeling being. From this point, as the most important of the whole, we come back to the presentation of the stick-laying.

We will now, dear reader, enter one of the Froebelian kindergartens. Here sit (since an interest in kindergarten has arisen) at one or two, or two times two tables joined together lengthwise (in pairs), and not so very far from one another, ten, twenty, up to forty children. The children greet us by joyously rising and turning toward us with their song of greeting:

> "We greet you, we greet you,
> Kindly we say,
> Welcome to-day,
> Welcome, welcome."

Kindergartner. "The children are quite right. You have come to us just at the right time. I might say with the greeting song (*Willkommen*), you come in accordance with my will—that is, my wish; and certainly also in accordance with the wish of all the dear children [the children, "Yes, yes"], for we are just beginning a new play."

I. "That is fine.—Do sit down at once, dear reader, here at this table; there is some space left."

Kindergartner. "I only fear that the play will seem too insignificant and simple for these dear visitors, for we are just beginning the play with the sticks, and especially the stick-laying. When you came in I was about to begin with a single stick."

I. "That is indeed very fine. I have actually, as you said, come just at the right time with my somewhat doubting guest; for all that is great, if one traces it back to its germ and to the first intimations of it, begins almost always with that which is quite insignificant, and the manifold goes forth from the simple, indeed the heavenly from the earthly, just because the latter contains the heavenly in itself. I always think of this when my boy in the company of his playmates cuts his reed flute in the spring, and when I go into the church and hear a Thuringian chorister of the genuine stamp playing on the organ, with its *vox humana* stop, the introduction to an "Allein Gott" ("One only God"), and then this choral itself, and see as an altarpiece the picture of St. Cecilia, the inventor of the organ, transported to heaven. This is what many of the dear visitors whom I bring here with me will not at all believe, viz., that the childlike, pure, and simple in its constantly continued cultivation should lead to heavenly glorification. This is the reason that the knowledge of the continuous, which has just been manifoldly illustrated by the reed flute and the organ, is so highly important. A visit to the kindergarten will, of course, not show us this to-day, but still it is a beginning. I am only sorry, dear kindergartner, that I have delayed you so long. Do go on, and do not let yourself be prevented by the seeming insignificance of the play from freely carrying it out in our presence."

Kindergartner.

"Gayly, children, one, two, three,
Joyous each in play will be.

What have I in my hand?"
Children. "A little stick."
"What can you tell me about the stick?"
Children. "It is straight, it is long."
"Do you know any other things that are straight and long?"
Children. "Yes; the—the——"
"Now only *think* of what you know; afterward you shall show it to us. What does my stick do now?" The kindergartner places the stick upright on the surface of the table.

Children. " It stands! "

" Do you know any other things that are straight and stand? "

Children. " Yes; the—the——"

" Now I told you before that if you would only think about it for a while you should show it to us."

The children look at one another and laugh, for they can not show anything.

" What does my stick do now? " The kindergartner lays it flat on the table.

Children. " It lies down."

" Do you know of several other things which are straight and lie down? "

Children. " Yes, yes, yes."

Kindergartner. " As you have answered me so readily, pleasantly, and quickly, each of you shall now also have a beautiful, new, clean, smooth, shining stick with which you will like to play." The sticks are given a few at a time to certain children, who then pass them quickly to the right and left as well as opposite.

" Now each of you can tell and show me of what your stick is a picture to you, of what it reminds you, what you think about it, and what you can imagine about it.

" I see in my stick my bodkin." As the kindergartner says this she lays her stick on the table, touching the middle line (|), and says several times: " A bodkin. What did I see in this stick? "

All. " A bodkin."

To the first child, " And you? "

" A darning needle."

The kindergartner lays the second stick beside the first at a little distance (| |) and repeats, " A darning needle."

" What do you see in the stick? "

All. " A darning needle."

So the kindergartner goes on till each of the twenty children at the double table has seen or shown an object in his or her stick, and the kindergartner has laid a stick on the table for each one, so that now more than twenty sticks lay on the table in the middle:

| | | | | | | | | | | | | | | | | | | | | | |

" Now we will see if we can remember all that each of you has shown us. Each of you must pay attention so that your stick may not be called by the wrong name or be wholly forgotten."

The kindergartner begins, pointing at the name of each object to the stick which represents it, and the children repeat:

A bodkin—a darning needle—a match—a slate pencil—a beater—a ruler—a lead pencil—a flower stick—a stick of wood—a sail needle—a netting needle—a cane—a yardstick—a candlestick—a candle—an I—a 1—a toothpick—a rafter—a pillar—one side of a ladder—a cigar—a penholder.

"Before, when I said that each of you should show me what you saw, you looked at each other, laughing, and thought you had nothing at all to show, and see now you have shown me more than twenty things.

"Now can some of you show me where each one's object lies?"

Children. "Yes, yes, yes——"

"Who has the match?" "I." "Where does it lie?" "Here."

"Who has the side of a ladder?" "I." "And where does it lie?" "Here." Etc., etc.

And now you, dear reader, who have accompanied me hither, have the confirmation, if you believe in, confide in, and see through the statement, and you who doubt and are unbelieving have the proof that manifoldness proceeds from the individual and single in the kindergarten; indeed, I may say that something proceeds from nothing. For instance, each child had at first scarcely one, or at most a few, images and conceptions, but certainly not one that was clear. Now each child has at least twenty, and certainly several more which were not only aroused within him, but were also made outwardly visible and named. Have you, who accompany me and examine so critically, observed how gayly and joyously all the children were intent on what each could see and show in its stick, and that each does not see and show the same which others have already seen in theirs? Do you see and note how the children look at one another with such joyous eyes? They now have a common treasure among them. By this common treasure there has been formed in and among them a quite peculiar spiritual bond of reciprocal respect and acknowledgment, and it would have been easy to increase these twenty perceptions by more than twenty others. In what different directions, and to what different sides in life, is the child led thereby!

Now, are not the children right, and do they not know and feel what they say and sing when they leave the kindergarten with the song:

> "In all things that we do
> The good comes into view,
> The beautiful, the true.
> Our playtime now is done
> And gayly home we'll run."

Now, dear reader, we also will go home and critically consider somewhat further what we have heard and seen, for I will freely confess to you that it has been with me to-day as it always is when I go into a kindergarten, that I come away from it each time as a scholar, even in my old age. Thus I have also learned something to-day. And what have I learned? May I tell you, dear reader? I have discovered, first, that the powers of memory and imagination are considerably increased by this employment, and long-vanished images come forth again from life in perfect freshness and connection. Second, that the power of comparison is exercised, things of the same kind are unconsciously brought together in the child's mind. Third, that the province of conceptions and perceptions is extended. Fourth, and finally, that the power of perception and comprehension is sharpened.

But who has not liked to learn something new and liked it so much the more as his knowledge of the disclosures of life springing forth from within increases? Manifold material for new perceptions forms the charm, indeed the magic, which, rejuvenating and vivifying, enchains one ever anew in the kindergarten—that is, if it be a genuine one.—
"May I, kindergartner, come back to-morrow at this hour?"

Kindergartner. "The master is always welcome, for in companionship with him one attains mastership."

"Will you, then, accompany me again, dear reader?"

"I think so."

Good-day, my dear reader! You have come early to my house to accompany me to our sensible kindergartner and her happy children. This is to me a joyful sign of your very lively interest in this simple subject.

"How could it be otherwise, for that is just

the peculiarity of the attractive subject, that it can be so enchaining in spite of its simplicity. For I will freely confess that since I left our active children in the happy kindergarten yesterday the subject has constantly occupied me, and I have actually regarded the things around me with quite different eyes. Indeed, the objects themselves have, as it were, drawn me back to it."

If you, my dear reader, are already so struck by this when you occupy yourself for the first time with the subject, how much more will it surprise you when I, before whom it has been brought more than a hundred times, am obliged to confess that this has been the case with me also since yesterday. I have been continually led to the sticks we had yesterday by objects which suggest the linear. Can you give me a reason or a comparison for this?

"I can not give a reason at this instant, but I will give you a comparison. The sticks attract as the magnet does the iron, or as the earth magnet attracts the magnetic needle."

Dear reader, you who thoughtfully and observantly accompany me, have given at once the key for the solution of our question by this comparison. In the innermost depths of the earth, unattainable by us, rests a power the nature of which we can not understand. This is the magnetic power which directly influences every magnetic needle that is so poised as to admit of free movement. This is one fact to the perception of which this comparison leads us. The second fact is that each magnet, when developed and impressed and, I might say, organized with magnetic power, is in a condition to arouse the magnetically attracting but as yet slumbering power in each simple and rough bit of iron, as well as to affect one which is similarly developed. Let us now, while in the kindergarten, contemplatively hold fast this threefold fact and expression of the magnetic power.

Thus, first, we notice the fact that the magnetic power, hidden in the impenetrable depths of the earth, acts in inscrutable and equally hidden ways on the developed needle unobstructed in its movements. The second fact we notice is that this developed magnet, as well as the hidden earth magnet, acts on the magnetic needle. But we especially remark the third fact, that pieces of iron and iron things—as yet undeveloped for the magnetism of the earth—call forth, awaken, and develop the magnetic power; at first, indeed, only passingly, but later, with continued influence, abidingly.

But now, to explain the attraction of our stick, let us hold fast merely the generalization of these three facts:

1. That deeply hidden, inscrutable, general power acts arousingly on already developed individual power;

2. That developed individual power acts arousingly, in like manner, on already developed power; and, finally,

3. That undeveloped power yet slumbering is indeed aroused by a power already clearly organized, but sinks back into itself if that power has not been aroused to a certain fixed point.

Here, my dear, reflective, accompanying reader, you have an intimation of the explanation of the magnetic effect of our stick, of the stick-laying and stick-play, and, at the same time, a hint of the explanation of the attractive power of each play in our play-whole, and of the developing, educating, actually formative power of this whole. The explanation is as follows: The magnetic force abides hidden in the innermost depths of the earth and acts with outwardly operative power on the smallest outside specimen of its kind which is developed to independence, and with this acts in the threefold way mentioned. In a similar manner a constantly invisible and not less attractive outwardly acting inner power abides in Nature as a whole, as well as in the parts of Nature. An example of this is given in our little simple and yet tangible stick, not only by the middle line which is indicated by its visible ends, and is hence, as it were, visibly invisible, but also by the middle of this line which is purely invisible, merely perceptible to the intellect, though never to be made visible.

But this power abides in the stick not merely as a part, but as an actual whole in itself, and at the same time a part of the great whole of Nature and life, here, first of all, of the earth. This inner power, abiding in, and wholly inseparable from, each visible thing around us, even if it be the smallest grain of sand or the bit of wood with which we play, this inner power hidden in each thing is what effects a certain invisible (if it is not permissible to say spiritual) relationship of things to one another, and also affects our cultivated minds by its arousing power of attraction as the earth magnet does the easily movable magnetic needle. In this extremely delicate yet effective reciprocal relation of the invisible, the innermost, in each thing to the things around it as well as to the observant spirit, lies the peculiar charm of the employment with the sticks, not only for our children but also for the kindergartners, and for every one who takes a true, warm interest in these employments. Indeed, whoever will give himself up calmly and without prejudice to what the activity of his own spirit requires will be unconsciously drawn into this sympathy. This is the effect of the hidden, always invisible, inner power. In this is conditioned the innermost nature and relation of the kindergartner to the children intrusted to her charge, and that of the children to one another, as well as, in general, the effect of education. But this fact has been hitherto wholly overlooked. Herein is also given the innermost nature of the kindergartens, as well as of the reciprocal relation between the kindergartens and the children who attend them, and also in general of that training of the human being which educates by developing. The nature of the power and of the spirit is here brought back to its most general expression and to its most general comprehension and perception, being the innermost. I place great value on this fact. We will later return to this subject.

But we have reached our goal. The song tells us so. Let us enter.

Kindergartner. "I have just introduced our little playmate at the present time, our stick, into our circle and welcomed it again. If you please I will go on."

"That is not only a matter of course, but it is our wish and the reason of our return."

The leader of the play, speaking in a singing tone:
> "One, two, three,
> Each must now attentive be."

"How many sticks did we lay at a time yesterday?"

"Only one."

"Well, each of you has a stick. But shall we not go further to-day?"

She is answered partly by quiet, bright, affectionate glances, and partly by a joyous "Yes."

"Well, then, each of you can have another stick. But tell me how many sticks each has?"

All together. "Two sticks."

"Now see once more which of the things you know you can lay with them; that will be a kind of drawing."

Shall we, my dear accompanying reader, join the children, ask the dear kindergartner for two sticks, and see if we also, like our dear children, can not make or draw something with the sticks which might not immediately occur to the children?

Kindergartner. "I do that also, as you, my dear visitors, will immediately see, as I am thus a child with the children. This has a very good effect on them. First, it pleases them very much when a grown-up person, and especially a dear visitor, will become a child with children. They feel more grown-up for this union. It rouses their mettle, increases their power of will and action, and their joyousness in representation. Second, new perceptions will be called out in them as older ones are recalled to their remembrance."

You see here in another way, my attentive, accompanying reader, the influence of the invisible, especially in the kindergarten. As the invisible is efficient in its working in the flower and vegetable garden, so is it here in the kindergarten for the good of the children who are guided and exercised by the conscious child-fostering.

As no sun ray, no shadow, is without its effect in a well-tended garden of plants, so in the genuine kindergarten no activity, even the smallest, is without its effect on the likewise receptive little human plants within. In this all-sided thoughtful consideration of that which is invisible, but just for that reason efficacious, lies the result of kindergarten fostering, often so inexplicable to those who have not advanced in the study of the science.

Kindergartner. " Now tell me what pretty things you have each laid."

A candlestick; knife and fork; a bridge over a brook (two sticks close together); the two sides of a ladder (two sticks a little apart from one another and parallel); a mill flume, or mill brook (two sticks a little farther apart); a hammer; an auger; a pair of tongs (two sticks crossing one another somewhat above the middle); a pair of scissors for cutting paper (two sticks crossing one another somewhat below the middle).

The kindergartner, as before, lays each of the objects represented by the children on the table in the middle, so that, first, each child sees itself recognized in its work; then, that the product of the individual may become a common good; and also, finally, that, for the future, the creative power of each child may be enhanced and its natural qualities, ideas, and conceptions be increased and cleared by the comparison of the different objects.

"But now pay attention to what I will show you, children. What do I do now?"

"You beckon."

"With what do I beckon?"

"With your finger."

"What is the finger with which I beckon called?"

"The pointing finger." (In many localities it is also called the beckoning finger.)

"Why is this finger also called a pointing finger?"

"You point with it."

"How do I hold my finger when I point with it?"

"Quite straight."

"What do I now do again with my finger?"

"You beckon."

"How do I hold my finger when I beckon with it? Is it still straight?"

"No, it is bent."

Now see here. The space within my finger when I bend it so is called the angle (*Winkel*), because I beckon (*Winke*) with it. See, now I place the point of the stick exactly in the angle of the finger. Where does my stick lie now?"

"Inside of the angle."

STICK-LAYING.

"See here, now I will beckon again. If I beckon again with my finger and make an angle with it, into how many parts do I divide my finger in that way?"

"Into two parts."

"You are right; and see, the lower part is still and the upper part moves. But see here again. Do I now make different angles with my finger when I beckon? How many kinds of angles do I make?"

"Three kinds: one is little and one is big, and one is neither little nor big."

"Now, because you have to-day laid such pretty things, I will tell you something more. The little angle is called a sharp angle; the big angle is called a blunt angle; and this one, which never varies, the right angle. Have you also a pointing finger? Can you also make an angle? Then make a sharp angle. That is right. Now a right angle, and now a blunt angle. But see here; I can also lay an angle with two sticks. What kind of an angle have I laid?"

"A sharp angle."

"Can you each lay a sharp angle with your two sticks?"

"Oh, yes!"

"Then do it. Quite right. But I can also lay a blunt angle. Whoever can, may do it."

"I!" "I!" "I!"

"But I can also lay a right angle. Who can do that?"

"I!" "I!" "I!"

"Good; lay a sharp angle. See here. The sharp angle, because it gets sharper or smaller all the time, says:

"I smaller grow, you see;

and the blunt angle, because it becomes always more blunt and larger, says:

"Wider to grow suits me;

and the right angle, because it always stays the same, says:

"I never do grow smaller, nor larger do I grow;
The same unchanging figure is what I always show."

"Now, dear kindergartner, sing the little angle song once more for us, and we will lay the angles." The kindergartner sings the little song mentioned by the children while they lay the different angles.

Kindergartner. "Who can lay something else?"

A collier's hut; a pair of compasses; a bridge; the gable of a roof; two turnstiles; half a sawhorse; part of a garden fence.

Kindergartner (to the visitors). "But will you not ask me something, so that you and my dear children may see if I also know something?"

"Very well. What do all the things which

have just been named show in respect to their angles?"

Kindergartner. "The sticks of the collier's hut and those of the compasses form sharp angles; the sticks of the bridge form a blunt angle, and those of the gable form a right angle; those of the turnstile form four right angles, and those of the half of a sawhorse and those of the part of a garden fence form two sharp and two blunt angles."

"Is your auntie right, children?"

"Indeed she is right."

Kindergartner. "Now, because you are so sure that I was right, I will show you something *right*. Stand up, all of you. Place yourselves straight before the table. Take a stick with the tips of the pointing finger and thumb of your right hand, hold it high above the table so that the lower end of the stick points straight toward the table; after I have counted one, two, three, etc., let it fall, and see what else it will do. Now pay attention:

"One, two, three,
Let the sticks go free."

All the children do it.

"What have you noticed about the stick?"

"When it fell down it sprang up again."

"Yes.

Upright now down falls the stick;
Upright then it springs back quick
In the air.

Now we will do it once more, and at the same time sing the song for it." They do so. "Now take two sticks in your right hand just as I do and tell me of what do these two sticks so held and so moved remind you. What object do they recall to your mind?"

Several. " A steelyard."

"You see the lower stick lies exactly right (*recht*), like the beam of a steelyard (*Wage*), so this position, or direction, is called horizontal (*wagrecht*), and the other stick, because it sinks (*senkt*) straight to the table in a right line, is called perpendicular (*senkrecht*). Now we will lay the sticks of the steelyard vertically and horizontally on the table."

REMARK.—In the actual kindergarten the perception and knowledge of the angles and of their nature is much exercised by the earlier plays and perceptions—by the limb, ball, movement, and building plays—and the knowledge of these angles in the kindergarten is here presupposed. Only, for the benefit of those readers to whom these plays (or at least the originally developed course of exercises in these plays) are strange and for those who wish to use the stick-laying without the preparatory plays, means and ways to develop the conception of the angle and its relation in the province of stick-laying should be here shown. Another reason why these ways and means should be shown is for the purpose of justifying the earlier expressed assertion that each play is in itself a whole, as well as a member of the whole—that is, bears in itself all the essential properties of the general whole, and so also of each individual play, but in its own peculiar way. A twofold phenomenon is grounded on this in and by the free creative employment of children: first, that the child in each thing, as an object of his activity, finds something (properties) which he delights to perceive and recognize; second, that the child treats everything as material with which to employ himself, by means of which to create and by means of which he can represent externally something which exists in himself. And so everything (especially the plays of the play-whole) stands (in a way of which the educator is clearly conscious) in a double relation to the child: first, in an arousing, representing relation; then again in a relation of taking up into itself, as it were, and re-presenting the child's conceptions and the activity resulting from them. In this double relation of our play material (well weighed in respect to aim and means) to the child, in the relation of the inworking charm and the outworking, formative impulse which stands visibly before the child as a result, lies the satisfactoriness for the child (once more through the play in general), as well as, quite predominantly, the joyousness in the employment with the plays of our whole of plays and employments presented with consciousness of the aim and means.

These inner, spiritual, invisible phenomena in the child, easily perceptible by the quietly and thoughtfully observing spirit, must be clearly recognized in order to recognize and correctly estimate the true nature of a kindergarten, as well as of the fostering of children and manner of employment therein practiced, and the spirit, nature, and effect of each play on the child. For this reason I have so long dwelt on the inner aspect of the play material, and especially of my objects of play, as well as that of the child. This will also explain (for on this the fact is founded) why already prepared play material is so much less liked by the children than the simple and little prepared, and why children are so ready to put aside the former for the simple material offered them in the kindergarten. Of this fact many examples have been related to me.

This deeper and more spiritual insight into the inner reciprocal relation between the child (who is unfolding himself in play) and his play material, was here required for the correct comprehension of the whole nature of that early fostering of childhood which educates by developing, which keeps the children busy, and which itself creates. But you, my dear reader, you, dear mothers, and you who help the mothers in many ways, as well as you, brave teachers of national and country schools, who lack time to obtain this clear insight into the inner nature of this way of employing children before actively engaging in your vocation, need not be discouraged from acquiring this insight. Do not let yourself be prevented by the above-stated requirement from constantly using this or any other play for the children committed to your charge, as the stage of their development requires. Only go according to your own and the child's natural impulse (the instinctive demand of your own as well as the child's inner nature), which transmutes itself in reciprocal charms. For you see the child's favorite playmate is the child who wholly gives himself up to the play and enters into it. Therefore, trusting in the child and the object of play, become a child with the child. Only in what you do, consider your employment, its grounds, influence, and results, and seek by degrees to become conscious of them, and you will, through your own trusting, examining, and comparing consideration of your own activity, attain to the deeper insight just required, not only into the nature of the child and of its life, but also into your own, as well as into the nature of life in general. This is just the blessing, the invisible reward, of that genuine developing nurture of childhood.

You, my dear reader, who were here like a child with the child and as busy as he, will have already experienced at least a part of what is here said. I rejoice in the quiet pressure of your hand and your grateful glance; these are an assurance of the benefit you will derive from coming here to-day with me.

Since you have followed thus far my presentation of the inner nature of this play, I must ask you to accompany me yet further. That which you see here before you consciously represented by the children and which shows a remarkable inner connection suggests to me the continuation of this inner connection. For the inner life of the children is revealed to us by what they have done, and we enter thereby the most secret workshop of their spiritual activity.

See, my kind companion, what lies here before us. Here is a turnstile (that is, one stick placed vertically and another placed horizontally and touching it in the middle); close by it is a turnstile in an oblique position (the two sticks crossing one another at right angles); near this again, two sticks crossing one another in the middle at an oblique angle in the extension of height; close to these is the same form in the extension of length. Is this indeed accident?

> If we look closely and compare,
> We surely find an answer there.

Now, then, what does our calm, comparing inspection show us? The diagonal or obliquely lying lines or directions which the two obliquely lying sticks show to us tangibly are intimated invisibly in the first sticks which stand straight and are at right angles; and, on the contrary, the right lines or directions which the two sticks which stand at right angles show us tangibly are intimated invisibly in the obliquely lying sticks. And so we see what the one shows visibly is in the other invisible, but perceptible to the inner eye. Here we see again the working of the invisible. Let us go further. Our eye connects the ends in each of the two crosses by four invisible straight lines, and so two squares result which lie opposite to each other (the first in an oblique position as an opposite square, the second as a principal square).* The two obliquely lying sticks which cross each other also show something similar; their ends, invisibly connected by lines, form also two quadrangles—not squares, but rectangles; the first, being in the extension of height, forms a high rectangle, but the second, being in the extension of length, forms a long rectangle.

You see, therefore, my thoughtful, accompanying reader, how here it is primarily the connection of the visible and invisible, the connection of the opposites, which is so attractive in the children's employments; but that which is invisible yet per-

---

* Referring to the perceptions gained by the fundamental folding.—Tr.

ceptible and which really acts is especially attractive, and therefore should be principally considered. This consideration constitutes the nature of a kindergarten.

Kindergartner. " Now, my dear visitors, you shall see what pretty things my diligent children have laid while you have been so eagerly talking to one another. Children, can you name all these things for our dear visitors? "

" Oh, yes! oh, yes! "

" Now, then," pointing.

One child alone. " A level."

Kindergartner. " Together."

All. " A level." In this way each object named by a single child is each time repeated in chorus in order to increase the perception.

" A smith's tongs " (the sticks cross each other below the two upper ends); " a long hop pole " (two sticks touching one another at the ends and placed in the same direction); " two bean poles "; " a lead pencil and a slate pencil "; " a whip "; " a carpenter's square "; " a sand clock or hour-glass "; " a fishing rod and line "; " a watchhouse in an orchard "; " a tent."

Kindergartner. " It is very profitable for the children to repeat the same forms with different names, as these forms then make a deeper impression on the children; for instance, the two last-named objects with the collier's hut previously made, or a church cross or churchyard cross and a signpost."

Kindergartner (to the children). "These two last things can not be well shown with two sticks. It will be better to leave them till you have three sticks apiece. That is true also of the level. Even my steelyard might have been left till then. Indeed, I should have left it till then if I had not wished to show you why we say horizontal (*wagrecht*) as I did before why we say vertical or perpendicular (*senkrecht*)."

Kindergartner (to the visitors). "I permit the children at times to make such representations as correspond but little to the actual forms, so that they may be led so much the more to look at what they have represented, and compare it with the object they wish to represent."

"Let us now count how many forms we have. Thirty-four. Then there are thirty-four forms and objects which you have together made. Now look at all of them closely once more. Now I will take up the sticks, and then each one of you may try to make as many of the forms you have just seen as you can remember. I will give each of you two sticks as often as you need them."

Kindergartner (to the visitors). "I like particularly to give this exercise, for it makes the children conscious of a number of conceptions and perceptions. This arouses the power as well as the mettle of the children, and also arouses the joyousness and activity of the representation, as you will soon see."

The children (now one, now another). "Please give me two more sticks." "Please give *me* two more sticks." "Please give me two more sticks, too."

See here, thoughtful reader, the effect of the formed and heightened power of life. Thus life early receives for the child a significance, the child's intellect receives a material for thoughtful comparison, its mind and heart the joyous feeling of satisfaction, its body the strengthening feeling of ability. The senses receive certainty of perception; the members, especially the hands and fingers, adroitness of representation. You can see here in this little insignificant play, my dear reader, the effect which is the aim of the whole of the kindergarten as well as of these plays as a whole and also as an individual part—fitness for life and life union.

Kindergartner. "I would like to call your attention to something else in respect to these two effects.—Emil, how many figures have you?" "Eight." "And you, Emmie?" "Ten." "And you, my little Maggie?" "Six." "And you, my diligent Robert?" "Fifteen." "That pleases me; it shows that you paid great attention. And you, my quiet Augustus?" "Thirteen." "That is right; always reach forth to the things which are before." "But I have eighteen!" "Who is I?" "Charles." "Well, you are the oldest and the largest, and so it is only right that you should have the most. You see, dear visitors, that the children have kept in mind different numbers of forms and different forms among those at which they previously looked. The cause of this difference is the difference of connecting power in the minds of the children; but partly in order to bring this power to consciousness and partly to strengthen and enhance it, I bring the different forms which have been again produced by one or more children into a little connected story. And if the objects are not too numerous I include them all. Thus it becomes apparent that each individual form has a purpose, and the whole acquires connection and significance. By this means also are attained by the child with consciousness what you just mentioned as the object of the whole—viz., harmony of life and fitness for life. This greatly delights the children, because their little creations thus receive life and recognition. The older children will soon try to find a certain connection between the things they make." More will be said on this subject at a later period.

What do you think now, dear reader?

"It seems to me that where such harmony originates in youth it must also be represented in later life."

"Farewell, children; be always as dear and good as we have found you to-day."

Kindergartner. "If you will wait a few minutes we will sing the reply.—Mary, Anna, Margaret, you may take up the sticks.—I let the children with whom I am particularly pleased help me in kindergarten as a reward." To the assembled children:

"One, two, three,
In a ring go we."

"Our dear visitors told you to 'be dear and good.' Now, what closing song will you sing in reply to this request?"

All. "Our playtime now is done."

Kindergartner. "Very well; we will go round to the right while singing:

"Our playtime now is done,
And gayly home we run."

The kindergartner with a clear voice:

"Farewell, farewell,
Be dear and good."

The children, softly answering:

"Farewell, farewell,
We will be dear and good."

To the kindergartner: "Farewell!"

V.

FRIEDRICH FROEBEL, HIS FUNDAMENTAL PRINCIPLES OF EDUCATION, HIS MEANS AND MODES OF EDUCATION, AS WELL AS HIS 'EDUCATIONAL AIM AND OBJECT, IN RELATION TO THE TENDENCIES AND REQUIREMENTS OF THE TIME.—REPRESENTED BY HIMSELF.

WHERE could there be now, in any position, vocation, or profession, a human being who had not been drawn by it to be observant of the different phenomena and manifold facts of the time in which we live?

But if we turn away from this multitude of phenomena of the time, hard and grasping as their effects may be on spirit, mind, and life, and direct our attention to the *character* of the time itself and endeavor to survey at a glance the whole in respect to its causes and foundation, to its innermost spirit and aim, we find that it is the impulse, the effort after development, cultivation, and continued training—in one word, it is education, the striving for education, which generally and vigorously moves men and nations, which gives to the time its character, its whole expression, which constitutes the spirit of the time and determines the aim of its efforts.

If we now seek out the final innermost cause of this it is expressed in an uninterrupted, lasting phenomenon which lies near to us—namely, in the revolution, the periodicity, and the cyclical recurrence of almost all the phenomena of life. What we now remark each day in the change of the times of day, each year in the change of the seasons, and in ourselves in the change of the ages of life, recurs also with the human race and humanity considered as a whole, as it were, as one human being. As it runs through the year in seasons, for example, it likewise necessarily and in strict order, indeed in unavoidable sequence, runs through great times and sections of its vivid expression, only that these changes are measured here by thousands of years, as in the former case by months, weeks, and days.

In the beginning of such a large, sharply defined section of time, such a period of the development of humanity and life, we now stand. We have lived in it for some time without having rendered to it the special and vigorous attention and observation which are really its due. It would be difficult to explain this neglect if it were not generally the case that that which is nearest to man (for example, the air, the light, the water), although most important to his existence, is often least noticed and considered by him, and greatly to his injury. In the same way but few observe the instant and exact time of the change of the day, of the year, and of other periodical and cyclical times, and thus perceive neither the end of the one nor the beginning of the other, although they are strictly dependent on hours and minutes, etc. But most people usually first observe the true advent of the new time, of the period of progress and development, after it has long taken place, is already long past. Such is also actually the case in and with the present time. We now first begin to recognize universally the advent of something altogether new; but even now we notice, though still imperfectly, what, alas, presses hardly upon us, how each neglect of development in accordance with the laws of Nature draws after it in life its natural, often painful, results.

The conception formed from natural history, and especially the astronomical cosmic conception of life and of the history of the human race, very greatly clear and enlighten life and the understanding of it, and we shall see immediately how this also expresses itself in the character of the present time and in its predominantly educational efforts.

But the whole life of man and humanity is a life of education. If we now reflect upon this the thought forces itself upon us, what is it now by which the present time especially proves itself to be a time of education, of the progressive training of the human race, of humanity?

Here, as the principal foundation for the satisfactory answering of the important question before us, appears, first, the fact that only *that* has real existence for man which has passed in and by clear consciousness, which, as it were, has been born anew in spirit, and indeed (again in the like conscious manner) was recognized not as merely isolated but as an active member of a greater whole.

But what concerns each individual, and consequently, above all, the individual human being, concerns also the whole human race, as has been already said. For the human race also, *that* only actually and truly exists which has in the greatest possible universality passed through the consciousness of all, and by which the human race is not only recognized and acknowledged as whole and individual, but again as part of a further composite higher whole.

Second, the fundamental answering of the above-stated question rests also on the fact that an era is distinguished only by that which comes forth not only in a few individuals but in an independent plurality. That which till now was actually educationally lived and carried out in life by individual men, and also indeed by individual nations, which was likewise felt by others, earnestly desired, and also by degrees raised to consciousness

EDUCATIONAL PRINCIPLES.

(according to possibility, or relatively) by individuals, is now to take a place in life, to be consciously observed by the minds and recognized by the spirits of a predominating plurality, as a spiritually united whole. Now whatever earlier times showed in respect to educational endeavors, as well in the individual as in the whole, they (the times) lacked either the universality or the clear consciousness of the endeavors.. But both in their association are the impressions of the younger or present time, and mark pre-eminently the entrance of a new period of the development of humanity and of the educational character of this period. But this is expressed especially and unanswerably in its individual requirements; these are as follows:

First, that the individual be pressed back into himself and led back to himself, whether this individual be an individual man, an individual people, or the whole human race.

Second, that therefore the human, instinctive, educating action determined by the higher tendencies of life be raised to clear consciousness, above all in the mother and the whole feminine sex.

Third, that the whole feminine sex be recognized by mind and spirit and actually acknowledged in life as a whole in its destination and dignity, as not only a real part and half of the human race, but also as being as essential to it as the masculine.

Fourth, that even the child and the life of childhood be recognized, acknowledged, and actually considered and treated in life relatively as a whole in its worth and dignity.

Fifth, the acknowledgment of the life of the family as a part of the life of the community, the reiteration and acknowledgment of the constant reciprocal relation between the two.

Sixth, it requires a clear and active co-ordination and co-operation of the social and political relation. As here of the more outward, so,

Seventh, of the more inward relations between school and home and of both to the church.

Eighth, clear demonstration of the relation of force, of mass (material), and of form to the idea, to thought; in general, to power.

Ninth, the endeavor for all-sided union of life with Nature, with humanity, and with God, and this is pre-eminently the undeniable proof of the real and prevailingly educational character of the present time. But on account of the great importance of the requirement of the time already brought out in this single enumeration, it is necessary to devote a special attention and consideration to it.

The first demand which thus characterizes the time in its educational efforts is that the human being be pressed back upon himself and led back to himself. This requirement appertains as well to the individual as to whole nations, and, it may be said, to the whole human race. Thus the pressure toward comprehension and consciousness of one's self (but also toward self-creation and personal action), which has long been felt, has given rise to the number of self-taught people (*Autodidacten*) and to the importance generally attached to knowledge acquired through personal action and personal experience, whether that action and that experience be outward or inward. But now this is quite evidently the effort of the human being, whether it be that of the individual, of the community, or of the whole human race to raise to consciousness, to insight, inspection, and oversight, the tendency to the activity and employment of life (the natural instinct). This is the effort for the acknowledgment of the human being in the triplicity of his nature as thinking, feeling, and acting in his united power of spirit, mind, and action.

But since now this must be done even according to the first and earliest fostering of the human being, thus even with the child and in childhood, and since this fostering is especially given by the Creator to the mother and in general to the feminine sex, and so to the family, the second demand which characterizes the present time as an educational one is that the treatment by the mother, determined by the high instinct of Nature and life, as well as the whole feminine influence, be lifted out of the instinctive as an influence which educates humanity, and this influence be raised to consciousness and thus to right recognition, to true acknowledgment. But the mother herself must, first of all, be raised to the recognition of her dignity and the importance of her striving.

But the third demand which characterizes the time as a genuine educational one is the effort to raise the feminine sex as a whole to recognition and acknowledgment of its destination and dignity and to living in accordance with the requirements of this destination and dignity, and especially to uplift its instinctive passive activity (as a part of humanity) and to rise to the same authority as the masculine sex on account of its nature and its vocation of fostering humanity, and so to consider the woman from her nature and spirit as purely and completely of equal birth with man.

But the fourth demand which no less proves the present time as an educating one is the effort to acknowledge the dignity of the child, of childhood, and of the life of childhood; not as single and isolated, but as a whole, complete within itself, as the germ and embryo of the development and representation of a life of humanity according to the words of Jesus, "of such is the kingdom of heaven," and so, I might say, to fulfill the fundamental demand of Jesus to his followers.

The fifth demand by which the present time is

proved to be a genuinely educational one is the acknowledgment of the family life (of the life of father, mother, and children, brothers and sisters) as a whole complete in itself and as the true root of genuine, pure, true human life. This acknowledgment is especially important for the family as a member (a branch of the race and stock) of the life of the community.

The sixth demand by which our time is especially characterized as an educational one is the reestablishment and acknowledgment of the reciprocal relation between the life of the family and that of the community, and of the fostering of that relation. Here it is again the relation of the individual to the common and united, of the especial to the general, which appears. If we weigh the importance of these single points against one another it is very difficult to distinguish which among them is of the most absolute importance, but, in reference directly to the present time, the one just named appears to be pre-eminent. It leads to the public spirit so necessary to us, lights up the relation of the one to the many, or, in other words, to the community, and leads thus to the regulation and arrangement of the social relations of the citizen, as well as of society at large. Hence the endeavor of the present time (by which it is characterized as educational) to regulate the relations of the community, of the city and of the state, as well as of society at large. For we know that the state in the totality of that which it gives or demands and takes is in a great measure actually an educational institution, whether good or bad we may not here inquire. Therefore,

Seventh, the present time is characterized as educational by the endeavor to fix the relation between home and school, and the relation of both to the church; or, properly, the relation between sensation, mind, thought, spirit, and active life; or really the question is of the relation between the inward feeling and recognition, and of the relation of both again to the outward action and to that which is outwardly created or effected. Thus the question is briefly of the clear relation of the inner to the outer, of thought to deed, of the idea to the reality, and on account of this effort,

Eighth, the present time is characterized as an educational one, just because thus the form as well as the mass, the material (the money), has lost its power, and, on the contrary, the real thought, the fine idea, the pure and good sentiment, in general, the spirit has risen to power; or also through the fact that the dead, quiescent, inherited possession is no longer a power, but the constant, spiritual, advancing cultivation which makes itself known in act and representation; so that in this creative action, in this work, it is the spirit and idea which animate it and give to the material the true value, often a hundredfold and more, as was long ago shown by Art. Or one may also say, the thought and idea attest themselves also externally as abiding, and yet again, as that which constantly renews and develops and rejuvenates itself from itself. They are thus become, as it were, genuine states; they now afford to life and to man what they require as such, and what it is the task and duty of the state to give; what man, in accordance with his nature, must strive to obtain. And onward development and onward cultivation of the individual by itself and as a member of the whole as such, with backward reference to the individual and to its needs for the pure living expression of humanity, is gained by endeavor; this is obtained when man is early treated as a creative being, as a creation of God.

From this, as the keystone and summing up of the whole, now follows that which points out the time as prevailingly educational. Thus,

Ninth, the general striving for union with life, Nature, and humanity, and consequently with God, which makes itself known in the most different religious and ecclesiastical efforts of the time. But at the foundation of them all lies the presentiment of the unity, the single foundation, the single fount of all existence, essence, and life. At the foundation of all is also the anticipation that man's vocation and destination lie only in undisturbed union with this unity, and thus in the purpose of real union with God, which gradually approaches its attainment. That is, they lie in a life in harmony with the laws of existence, development, and life (laws of life in appearance) which make themselves known in all beings as having their original cause in God and proceeding from God. Thus only in that consciousness of true union with God which attests itself immediately in action does man's vocation consist. This living expression of his nature, by his own choice and his own determination, consequently in freedom, effects the genuine peace and pure joy of the life of man, and is the total endeavor of the time as truly educational.

These are the most essential strivings which express themselves, not only clearly, but even audibly, in the events of the present time, and the most essential requirements of the new peroid of life, requirements which appear to us as unavoidable.

Yet it is by no means merely characterized as a predominantly educating time by the fact that these efforts are present, partly single or even partly connected, for they were present, isolated, or partly connected at different times, indeed at all times. No, the grandness, the high significance of the present time lies in the fact that *all these efforts are present at the same time*, and, it may be said,

EDUCATIONAL PRINCIPLES. 173

almost all in equal power and strength; which has not been the case in any other time in history. Then what is yet more important, and what points out this time in comparison with the former as exclusively educational, is, that *none* of these requirements can be satisfied *alone,* but that they must necessarily be *all* fulfilled at the same time, even though only *one* of them is to be completely fulfilled. And the present time makes this requirement for the education of the individual, and indeed for the education of the family and community, for the education of the people and of humanity, for the education of society up to the genuine State, that is, to the State constantly self-renewing.

But how is this apparently difficult requirement to be met? How is this important problem of life to be solved? Simply thus, that in the same manner as the gardener or farmer educates his plants to perfection in all-sided coherence with Nature and conformably with all requirements, *we* strive to observe, to develop, to educate, and to form the child, the human being, conformably to *his* nature, to *his* inner laws, and in untroubled union with life and Nature, in constant union with the origin of all life.

This manner of comprehending and treating the child and human being in the all-sided coherence of life proceeds (as being constantly the same result) from self-observation, from the observation of the nature of the human being and child, in general from the observation of all development and formation wherever it may show itself, and it thus expresses itself as the first and uppermost maxim, as the principal requirement of human education.

From this all-sided observation of life for necessary application to the education of the child, and thus in essential reference to the solution of the problem in question, we are met by the following highly important facts, viz.: that that which lies in a whole lies also in the smallest part of it; thus, that which lies in humanity as a whole also expresses itself even in the smallest and youngest of its children. And further, that thus, that which lies in humanity as a whole and expresses itself even in the child, slumbers in the child as essence and germ, makes itself known again in the smallest details of his nature: indeed, definitely shows itself therein to a clear, spiritual eye. This is the second maxim on which the method of genuine education of the human being and child is founded and through which consequently the problem in question farther receives its solution.

From this comprehension of the human being and of the children, from this comprehension of the individual life as well as of the life of others and of objects in general, proceeds farther the third essential principle of education, viz.: that as the inner development, the development from within,

EDUCATIONAL PRINCIPLES. 175

is joined to an impulse working out from within, so the outward form also depends on an attraction affecting it from without; and these two limitations, opposite to, yet like one another (for all life has indeed a single, thus the same, origin), give the single life uniting both in itself as a result (product)—the educated, cultivated human being.

We must therefore necessarily recognize a fourth law of development, education, and cultivation, viz.: that the child is truly formed to a man only by the co-working of limitations (factors) opposite to, yet like one another, and by the comparison and connection of these factors in and through life.

Conformably to all this, the first effort, as well as the first duty in the present time, must necessarily be for each one who acknowledges these demonstrated principles of education and cultivation to be true, to begin to apply and carry them out, first of all with himself, then with those committed to his care, but especially to begin to labor to introduce all educators, especially those of the feminine sex (first of wives and mothers, but also their perhaps already grown-up daughters, and educational helpers), into these principles. That is easy, because one needs only to clear, to strengthen their sure, natural instinct, then to raise this instinct to consciousness, and so to firm, logical, continual accomplishment, and to provide the necessary means for rightly following that which is understood.

But it would again be one-sided, if education for the training of the human being should be confined to the feminine sex, for the reason that women, who are faithful to nature, are the first trainers and educators of the human being. No! the more outwardly instructing masculine sex, according to the necessary law of opposites above mentioned, has no less a part in this training, as the future teacher, protector, trainer, as the future father of the family, of the community, and of the nation. This co-operation in the training of the human being must begin not only in the years of boyhood and youth, but even in those of childhood; so that the child may be early led on all sides toward his destination, in order that in the fulfillment of his destination he may protect and uphold the other, the softer, more delicate sex.

Therefore, in conformity with this whole system, humanity must be observed from the beginning of its dual existence as composed of two opposite sexes which are, in all other respects, alike. And so in its whole nature, its senses and limbs, its body and its soul, its feeling and being, its understanding and intellect, its comprehension and reason, and its all-exhausting nature, the child must be treated in accordance with the spirit of this system, consequently as a being rising from sensation to consciousness, to intelligent willing and doing, as a being one in itself, therefore sentient, reflective, moral, chaste, and having a high vocation.

Since, now, the germinating point and the source of all genuine development, of cultivation, and education is in the feeling and the sensation, as well as in the anticipation (therefore in the mind), this must necessarily early find its suitable nourishment, even with the first development of the child's body, limbs, senses, and spirit. This is done by introducing the child at once into the realm of harmony and accord, into the province of rhythm, melody, and dynamics, and thus into the realm of tone and song, for which the child early shows decided inclination.

We perceive how the nurses quiet the young child by the employment of rhythmical movements (of the so-called dancing, rocking, knocking, etc.), and the child actually feels itself pacified by these means. It appears even to agree with the pulse-beat, consequently with the heart, and, as is generally thought, with the seat of sensation. Thus we early see the healthy infant which has been satisfied in all its needs lull itself to sleep with singing when the mother has laid it in its little bed. Conformably to this expression of the child, and taking it up fosteringly, we hear not only how the nurses induce their charges to sleep by song, but also how they, whenever necessary, at other times quiet them with it. The somewhat later phenomena in child-life also prove that rhythm and song are intimately connected with the child's expressions of life.

Rhythmical, measured movements and harmonious song thus necessarily and early belong to an education of the human being, which, meeting the demand of his nature on all sides, is consequently a healthy one. So we must clearly recognize that the educational requirements of the time, earnestly presented in the foregoing remark, can be completely obtained only by the appropriate co-operation of song, even with the early education of the child, since it strengthens that which is noble in the youngest child, and makes himself more receptive of the noble. Hence the first plays of the body, limbs, and senses, the little plays of the Mother-Play and Nursery Songs practically produced for the earliest period of childhood and infancy, are mostly accompanied by song; by which at the same time the word, leading to comparing thought, is introduced (as most essential to human education, to the education of the human child) into the first strengthening and developing nurture of children, which is so important. This introduction of the word by song, which took place at first instinctively by motherly caressing, now takes place especially by the singing-tone, which exerts so much influence.

But with such development of senses, limbs, and

body, as well as soul, a development begun on all sides, penetrating mind and feeling, and consequently fostered (though only in the germ), the child now also needs an object by which he can develop himself yet farther and more completely, independently, and spontaneously. For we early see how, in order to use, strengthen, and exercise the power of his hands and fingers, he tries to grasp objects—his own cheeks, etc.; how he squeezes his own thumbs, holds fast the mother's finger which he has seized, etc.

The nature of the human being requires (as is early shown to us in the child) a corresponding counterpart for the animate being, for his inner nature, and for the requirements of that nature; that is, an antitype, opposite to yet like him. This requirement is one with the human being, inseparably connected with his earthly appearance and existence, and therefore repeatedly comes forth with the observation of his first appearance everywhere, in all zones, and in the most different relations.

If this requirement of the human being in general is not fulfilled for the child by a suitable object coming to him from without, he seeks to satisfy this requirement of his nature by means of his power of imagination (fancy). But the images of fancy lead the human being and even the child very easily into the boundless and formless, as they at the same time more weaken than strengthen the human being, and this even in his early development, at least for outward representation. They lead more to one-sidedness than to that all-sidedness of his cultivation, toward which the human being even in childhood is to strive according to his nature. An object must therefore be given to the child, not merely for his outward bodily activity, but rather for his inward activity, the activity of his soul, and for the development and cultivation of this activity. It is by no means unimportant, it is, on the contrary, a thing of the highest importance, what kind of an object is here provided for the child as a true counterpart of himself. We see also by what has just been said that it must neither be left to accident nor to the arbitrary will of any one what object is chosen for this purpose, but that this is already definitely given with the appearance of the human being as a child. According to the requirement it is to be an object like the child, but at the same time his pure opposite. Let us now observe and question the child himself, and see what object he chooses in his earliest period of development for such a counterpart of himself and of his efforts. It is the simplest inanimate object but also (a highly remarkable fact) the heaviest. He prefers wood and stones. The boys like to carry what is large and heavy, and seek it out as a plaything first of all. Little girls also make their favorite dolls of the heavy, large bootjack or a like piece of

EDUCATIONAL PRINCIPLES. 181

wood. I was informed by a mother from the higher family circles of the city, who was an early observer of life, that a heavy sandbag she accidentally found became to her the most cherished doll, when she was a little girl, because it had in it the weight of the actual child, and so she experienced an illusion and gave herself up to it; she imagined herself to be carrying a real child.

But weight, attraction, is the first expression of the power, as it were, the life in Nature which expresses itself in its higher degree as the attraction of the senses; and, in its purest development, as a spiritual attraction, as love.

In this course of development of the human being here pointed out as it makes itself known in each child is at once clearly expressed how the child is to be treated, what object (when? where? and how? why? or wherefore?) is to be provided for the child in the beginning of his own activity.

As a being complete in himself, bearing life in himself, developing and appropriating life to himself, and the opposite of life in his own adjusting nature, the child seeks also as a counterpart of himself an object which is opposite to, yet like himself. It must therefore, firstly, as a similar object, be such a one as will enable the child, for the free unfolding of his self-determined nature, to make from it everything which he wishes: that is, to conceive and think of everything in it.

Therefore, as purely opposite to the child, it must be, secondly, an object, a means for something else, while the child is himself only a pure end and aim for himself.

This statement clearly gives the character of the first plaything of the child. In this is clearly, anticipatingly expressed the deep sense with which the child chooses a stick, a stone, a board, a piece of wood, a bootjack, sandbag, or even the loam, the heaps of earth and sand, the clay and earth hills in which he can dig with zeal. Yet, after all, for his freest development, he likes best the ball. Just the ball is demonstrably the middle point and point of union, I may say the representative of all for which the child strives as a counterpart for his self-development and cultivation. It shows completeness in itself, and is yet the general representative of all things—of rest and movement, of totality and unity, of that which is all-sided and that which has but one surface. It unites in itself the visible and the invisible (its middle, its axis, etc.), By the ball, the child can now accomplish and represent unnumbered things which exist within him as desire, idea, and thought. And, with the ball, the child can imitate innumerable things which he sees around him. The ball is thus a means of representation for the inner world as well as a means of introduction into the outer world which surrounds him.

By this is solved the question, " Why is the ball as a plaything so dear to the child?" The play appears (corresponding to the sense of the word Spiel) to the human being, and especially to the child, as a mirror (Spiegel) of his inner and of his surrounding world, and is, especially at the stage of childhood, a mirror required from inward, therefore free, impulse by the child's attraction to life and employment. The plaything (Spielzeug) is thus (as a means and aim) opposite to, yet like, the child's nature; and for this reason it is the object which awakens and engenders (erzeugen) the desire for play, the act of playing (Spielen), and the contents of the play.

Play is thus actually engendered by the connection of opposites which are also alike, by the combination of the free activity of the child with the dependent movability of an object and its consequent power of taking form.

The ball is thus actually a gift which, in combination with the child's impulse to activity, by its many kinds of movability and its manifold employment, engenders the desire for play, for free reflection of the inner as well as the outer world, and is therefore a favorite gift.

If we now combine this statement with what has been before said, it is easy to explain why the ball is the child's first and dearest plaything, and why it remains so in the German ball plays through the whole age of boyhood and up to and into the age of youth; why it, as the best-loved plaything, pre-eminently captivates the youth who is early striving for all-sided cultivation.

That which especially concerns the play with the ball as the earliest and first means of development of the child can be done by the ball by itself and for itself in its simple form and in its simplest relation.

It can be used either free, or fastened to a string, and in both cases either in free space or (in reference to surfaces) perpendicularly, horizontally, or obliquely. When thus used the ball appears as a guide in the outer world by representing the objects of that world, as, for example, a kitten, etc., or as a cultivator of the child's own body and limbs; one may say, as an instructing gymnast.

The ball has been hitherto taken up merely in its form, and, in this, merely in its relation to the child and to the outer world. But it can also be considered and used as a plaything for the child in its relations to itself; firstly, in relation to its size; secondly, in relation to its clothing, its covering, or its color; thirdly, in relation to number; fourthly, in respect to its material; fifthly and finally, in its different relations of hardness or elasticity; and, in connection with both of these, by its falling on a corresponding surface in relation to tone.

EDUCATIONAL PRINCIPLES. 185

In all these respects the ball enters into ever new relations to the child, and the exhaustive treatment of it here shows it as a constant, all-sided, uninterrupted means of education and cultivation as well as of the actual development of the child.

A slight indication in regard to color must here suffice. The child, instinctively, and in accordance with his nature, soon seeks that which is the simplest opposite of the single, yet like it, viz., plurality. The fosterer of early childhood must, like a gardener, bring to the point of unfolding this unconscious effort of the child (his instinctive longing for plurality) and must also develop it to consciousness, that it may not become merely a greedy desire for possession and a strong longing for owning.

But mere plurality of a like kind, as such, does not and, according to his nature, should not satisfy the child, for he seeks in plurality the connecting unity or to have the manifoldness of the connecting bond made perceptible to him. This is most satisfactorily done by the colors of the balls, which are of like form, size, and material. The color should be the purest possible; six rainbow colors (seven by using a dark as well as a light blue), blue, green, yellow, orange, red and violet—the six (or seven [with indigo]) children of the light (in a prismatic spectrum) which is in itself single—which, as a wonderfully beautiful unity complete in itself, show in the rainbow the symbol of the highest peace, of peace between heaven and earth, between God and man. And why should not the path to such peace be early entered upon in a childlike way for the human being and for the child? Therefore, are brought before the child as a plaything, by degrees, six balls in the six designated colors, now singly, now in various connections. These balls, even singly, are received each time with pleasure by the child. Then they are given together in pairs as complementary colors: red and green, blue and orange, etc.; or combined in threes, for example, as the three principal colors, red, blue, and yellow; and as the three intermediate or mixed colors, violet, green, and orange. Therefore, here the combination of a different number with the conception of some particular shape can be early brought to the child by the simple and thus natural grouping (as it were, self-resultant) of two, three, four, or more balls in a closely connected whole. With the increasing age of the child increases also the size and hardness, and thus the elasticity, of the ball, as well as its capacity to call forth a sound by dropping it on a firm horizontal surface. We have already tarried a long time with the ball in order to show it to be necessarily the first plaything of the child. Yet that which is here stated about it, and, as it were, in its favor, is quite insignificant in respect to its being carried out in detail when compared with the genuinely educating and forming effect of what can be actually done by the ball, and, with it, by the child at each attained stage of development.

What has been here brought forward about the ball may perhaps seem too much, though it is but little in comparison with what the ball, as the first plaything of the child, contributes toward the little one's true and constantly developing cultivation. But we have intentionally lingered so long over this demonstration of the all-sided necessity, the indispensableness of choosing the ball as the child's first plaything corresponding to all. We have lingered intentionally over this confirmation of the fact that the ball is actually the true, first plaything of the child, because just this fact was so much disputed, at least the effort to again introduce the ball in its old right as the first plaything of the child was considered one-sided and strange.

And yet for the genuine educator, the ball is just as necessarily given as the first plaything for the child, as the spherical form of the earth is for the first step of the geographer; although educators of children have given every other object—but no ball—for a plaything, and although at the beginning of the knowledge of the earth many things were dreamed about the form of its upper surface, this being considered, now as a disk swimming on an endless surface of water, now as a disk supported by columns, etc. Indeed the ball is just as absolutely given as the first plaything for the human satisfactorily developing child to those who truly know the human being, as the spherical form of the world is to the satisfactory insight into the system of the world to those who are learned in regard to the universe; although the heavenly bodies have been in the beginning by no means correctly considered merely as lights, etc.

The fact that the ball belongs to the first stage of the child's development is also proved in many parts of Germany by the natural disposition of the country women to return from the market with a half-penny ball as a gift and joy for the little ones, though it be filled only with sawdust or cow's hairs. Another proof is the fact already mentioned that in many of the countries of Germany the play with the ball in manifold ways and with various alterations appears to be the favorite amusement of the smallest children, as well as of the growing-up boys and girls, even up to the age of youth. A Persian legend even indicates the ball as the privileged play of the king's children.

But the ball is highly important from the intellectual, sentient, and moral side, as well as from the corporeal and thinking, as a moral means of preservation, as a talisman; since, by the provision of the ball for free and full use, the child is preserved from ill-humor, and from all the moral diseases which proceed from it. The ball has the same influence in reference to the passions and emotions, since the ball neither arouses nor nourishes them, but, on the contrary, strengthens the impulse of the child to the activity and employment corresponding to its nature, develops in harmony with the laws of his own life, and leads to formation.

But enough has been said, I hope, for the all-sided confirmation of the choice, and for the satisfactory justification of the use of the ball as the first plaything for the "joy" of the child.

The total activity of the child proves (and this became prominent even in the preceding remarks) that he advances according to his nature, his life, and the law which expresses itself in his life (thus with general necessity and with especial joyousness and freedom) from that which is given to that which is opposite to, yet like it. This is implied in the common saying, "A child always wants something new."

But that which is opposite to yet like the soft ball is the hard, firm sphere which is therefore, according to the simple and natural course of development, the plaything next required by the child.

The sphere is more complete than the ball, and is also more easily movable, as its surface is smoother; but it is also heavier, and therefore rests more firmly and determinately when it once rests. Yet, on account of its greater weight, the child's use of it makes demands upon his more developed strength and dexterity. But, at the same time, that it makes more demands on the more developed strength and dexterity, it also shows by the noisier sound caused by its use, first, its greater weight, and second, the greater strength required to handle it: both of which please the child as the expression and proof of his increased power in the play with the sphere.

All this shows plainly the fact that this progress in the material proffered for play, and consequently for development and cultivation, is, in several respects, in accordance with Nature; and this is highly important. In the same way we certainly unequivocally perceive what is already expressed by the just employed phrase, "in accordance with Nature"—namely, that the advance is not an arbitrary one, but is given of necessity, since it includes in itself likeness and progress as well as contrast and stability, and therefore the indispensable conditions for such a progress of the child and human being as will be at the same time wholly satisfactory and worthy of humanity. These indispensable conditions are contained and expressed in what has been before brought forward. In the same way the condition of the development of each thing and being shows itself as necessarily given at the same time with the thing. This condition is, that what is to be developed, drawn forth, brought out in the later must lie as a germ in the earlier.

This is one of the most important of all the

general laws of human education, but has been till now but little noticed.

It is highly remarkable (and it is here made prominent once for all in its importance) how in the conviction of the nature of the human being here brought forward and represented, the requirements of education, as well as the means and ways of education, reciprocally limit and explain, justify and confirm each other. This takes place inwardly and (without being sought for) with an unmistakable necessity in the laws of development given with them and the way and method of treating and educating the child proceeding from them. This is to us the innermost and deepest proof of the truth of the whole, which can be grounded only in the perfect comprehension of the whole being and can proceed only from that comprehension.

According to the inner condition just presented, the sphere now appears as the second playmate of the ball. The logical deduction from the preceding remarks is, that the sphere by no means supplants the ball, but, with its aid, effects the farther development of the child who loves them both.

On account of its by no means slight importance, I here bring forward once for all, the fact, that the playthings or means of play, the following play-gifts, as we call them, never preclude the employment of the preceding, but that the use of the one is only yet more extended, explained, etc., by that of the other. The use of the sphere in play and the features of the play, or, in other words, the employments with the sphere as opposite to, yet like the ball, have naturally very much in common with the play with the ball, only that the movements of the sphere are much more exact and defined than those of the ball. A loop of wire, to which a string may be fastened, is affixed to the sphere, so that it may be moved by the string.

Each plaything is, in a certain point of view, a complete one (as, for example, each of the senses of the human being is itself a unit, and the senses collectively form again a whole of a higher kind). So each plaything has its appointed task to accomplish in the development and education of the child to the stage of maturity, and this task is to be accomplished by means of this development and education. As now the ball is to lead to harmony and accord, particularly by the variety of its colors, so the sphere is to lead to the clear perception, comprehension, and retaining of unity as such, especially in and during the variety of its turnings and twistings, by which, however, it clearly and unalterably shows the one sphere.

This clear and precise perception of unity in life through all its changes *may be*, and the quiet holding fast of this perception *is*, a quality which we all need, that we may preserve the peace of the heart, that we may attain to the joyousness of life, and that we may secure firmness of character in all conditions and under the most different relations. To lead the child to this in the most gradual and playful way, early to guide a number of children to this and to confirm them in it, will bring the blessing of genuine education into several families at least, and perhaps into a community, or even into a city or province. For the proof of the godlike in life and of the influence in life which is grounded in the godlike is that the blessings of it lead back to the smallest, proceed from the smallest, and yet stretch out far and wide.

The sphere is intended to benefit the child by developing his power of perception and conception and even his character, in play and by means of play, though quite unanticipated by him. It is also to develop the body and its members as a gymnastic model, as it were (we used this significant expression in speaking of the ball), by its manifold turnings and twistings. In order to avoid repetition for all subsequent play, I would here state that words spoken and sung (consequently also verses and little rhymed songs) in a manner corresponding to the child's state of development, to his head and heart, his thought and feeling, his mind and spirit, are to make a reality of the early entrance upon the path of education of the human being to all-sided development of himself.

Be it here remarked as essential in respect to the sphere and the play with it, that, as with the ball, the colors (as it were, like the joys of life) form a symbol of plurality, so do also with the sphere, the light and shade, the day and night sides of life. (As white and black spheres form, as it were, the opposite poles of the color circle which resolves itself toward one pole into the light, the white; toward the other pole condenses into the dark, the black.)

This small, almost insignificant, alteration gives to the plays with the sphere a great charm for the child, and very rich application to actual life in respect to its most different sides, especially with a grouping of many children; so that the use of it for the development of the practical employment of the children of all conditions must be clear to the unprejudiced eye. By means of the new, added shades, the relations of number, form, and rhythm appear in as new a light as a beautiful country does by means of corresponding shades of light when thrown on a white ground.

What is now to be the indispensably necessary advance to the next plaything?

The sphere has one surface, which is therefore a curved one. The contrast must have straight surfaces and several of them. The sphere has no corners and no edges; the contrast must have corners and edges. These are the opposite properties which the next solid used for play must show. Now for the similarity: The sphere has three similar directions or axes, reciprocally intersecting each other at right angles, and these axes must appear clearly and precisely when the body is at rest. The next solid used for play must necessarily have these like properties together with the above-named opposite ones. But this can only be the cube or hexahedron; therefore the cube is with indispensable necessity the third developing, educating playmate of the child.

On account of the plurality of its properties the cube, in comparison with the simple, round sphere, shows and gives a plurality of use and a multiplicity of the most different appearances, as the sphere shows an all-sidedness of movement. Thus the always stable cube, with its straight surfaces, represents itself to the child as the opposite of the round, easily movable sphere, but yet similar to it.

So the cube first shows to the child by its surfaces, corners, and edges, the purest contrast of the all-sided extension in one plane by the surfaces, and the all-sided convergence to a point by the corners. It shows also, by the edges, the connection of the two in the line, as it were, since they can stretch out indefinitely on two sides, but on the other are drawn together like the point. The law of connection was already approached with the ball in the colors. This law now shows itself to the child almost constantly in each of his playful activities, and so as an all-prevailing law of formation and life. It is later essential for insight and recognition, as well as for creation and action; for arousing and fostering the moral part of man's nature, as well as for all-sided, purest, and highest life union. It is important that this law be now brought to childish simple notice and perception in a childlike way even at an early stage of the child's development. The necessity of this requirement and of quiet obedience to this requirement very soon reveals itself for the welfare and pleasure of pupil and educator, of trained and trainer. The child's first pure incitement to comprehend and carry out all that is great and good in life is his pleasure in so doing.

But in yet another respect the perception and contemplation of a comprehensive law of life and development are important even in the earliest education, since we may not forget that we are perceptively intellectual beings, and that our first education especially requires the corresponding perceptively intellectual contemplation in order thereby to rise to a purer, more intellectual perception, and, finally, to inner spiritual comprehension which must be as free as possible. Up to the present time this forgetfulness has, alas, been shown, to our great detriment, in many ways, especially in our earliest primary school and national education, the sad result of which are now evident. As genuine, beneficially acting educators, recognizing the deficiencies of our present training and called to improve it, we must above all give again the genuine and comprehensive symbol to the education of our children and youth, and the many-sided education of the people, based upon the first education of childhood and youth. This only can furnish to our people what they need, for just the empty, effete ideas which have been committed to memory in certain logical connections have made the people also empty and dead, and weakened them for vigorous comprehension of the right. Here is a principal cause of the perplexities of life, for which reason I felt myself imperatively urged to linger so long on this part of my subject.

We now return to the cube as the third object for the child's play and development.

By its form it leads, *firstly*, to the perception of the solid form and to the knowledge of its boundaries, of the sides, the edges, and the corners (surfaces, lines, points), and of their different relations to one another in form, position, and size.

One must of course see for himself among the children the way and manner of introducing the cube, in order to convince himself that it is possible to do it in a manner corresponding to the child's nature and the then existing stage of his development, and that this is actually done. It is in many ways difficult if not impossible to present this merely by words.

Form, size, and number are important to the comprehension of the figure and to the perception of its interior, and are therefore important for life. But the clear, sure gaze is just as important, for this reason, that early notice of both inward and outward, and introduction into the perception of both, is a great gain for the whole life as a whole. But with the human being as a child everything begins in and with the comprehension of what is perceived by the senses. Therefore the early introduction of the child into the perception of form, size, and number lies in the nature of the human being and in the nature of the child, only not in the abstract, bodiless, and objectless, but connected with bodies and objects.

And so also the cube, in the same way that it leads into form and size, leads, *secondly*, into number and its relations, in the most constant natural way, which is therefore also the most agreeable and most judicious for the child. It thus appears like an entertaining teacher of arithmetic in the most manifold numerical connections of its sides, edges, points, angles, etc. But here we must again refer to that which concerns the truth of the whole thing, to that perception which is gained in child life, but is still more obtained by the *third* way of viewing the cube where it shows the most various and peculiar appearances in its different positions and movements. It is especially suited for play in respect to its manifoldness, which fascinates the child. It is also important for child, man, and life in its higher meaning. Such plays show and demonstrate that the human being is born for research; that he is to practice it even as a child, as also that he is, just as early in life, to separate that which *seems* from that which *is*.

Yet this is, of course, by no means all that can be said and brought forward of that which the simple cube develops from itself by activity and different ways of perception. For in the notice, etc., of form, size, and number, and of the different ways of appearing with one and the same fixed form, *fourthly*, it leads the child in a childlike manner into the fundamental ideas of physics and mechanics, the science of movement by its pressing on the hand, etc., and by its rising and sinking by means of the string. *Fifthly*, it serves as an introduction into life and to the objects of life by the different ways of perceiving and looking at it, which proceed from the child's fancy; for example, as a square stone, as a bale of goods, as a chopping block, as a tree tub in a greenhouse, etc.

A remark in respect to the playing and the nature of the play of the child here obtrudes itself. It should perhaps have been rendered prominent even earlier and must not remain longer unnoticed, because it is so highly important in respect to the child's whole occupation, to his relation as a perceptively intellectual being, to the object of his play and the use of that object, and also to his contentment. This remark is, that the child's satisfaction in playing and his delight in what he plays are by no means peculiarly connected with the outward appearance and value of the plaything, and with the perfection it shows, but rather with what the child can represent by means of this object, with what he can conceive, perceive, and imagine by means of that which is outwardly represented. The high importance of the child's playing and of the games he plays is that what takes place within him is awakened and developed during the play and, by means of the playing, takes form. It is this, not merely the object of play as such, which gives the child pleasure in his play, and causes him to be satisfied by it. Therefore the child likes best that plaything, whatever its outward appearance may be, by which and with which he can form and accomplish the most: that is, can call forth in himself the greatest number of and most satisfying conceptions, imaginations, and fancies as vividly as if he saw them actually in himself and outside of himself, even in the most imperfect outlines and representations. This perceiving and actual representing of things in the outside world—even though very imperfectly, yet always as a whole complete in itself—is just what gives such an exceedingly high, strengthening, as well as developing value to the representation plays as compared with the empty activity of forming abstract ideas or acting out such ideas. These representation games are plays of the fancy awakened by actual life and connected with it. But that occupation with abstract ideas can develop itself boundlessly (and so also into the formless), and yet be connected by no condition, no possibility, to the fixed, clear, and precise life-forms. However incomplete these representations may be in themselves, yet they are a self-contained, firmly defined, sharply bounded, already existing whole which can now be completed and perfected by continued cultivation, perseverance, diligence, dexterity, and skill: all of which can in a certain point of view be obtained by firm will.

Hence the actually quite incalculable, priceless value of the early exercises, the modes of play and employment, as, in general, of development, education, and cultivation of the child and youth which are here demonstrated and entered upon, and which are alike important for all conditions and relations of life, but especially important for all conditions and callings of practical and executive life. For they free man from the life of empty and formless, vacant, as well as measureless imagination and fancy, which is so inwardly full of disturbance, outwardly demolishing and annihilating, or at least perplexing, because they give to the human being and to his life all that his spirit hopes and anticipates, all for which his heart yearns, and which his outward existence requires. They give, on the one side, material and contents, substance and form, value and dignity to life, and thus to thought and feeling as well as to action. They give life, vocation, aim, and the determination of one's own destiny, to the human being, to the individual as well as to the whole human race. These means of employment indeed preserve throughout the happy and satisfying idea which is to become truth and reality, and which actual life gives.

" I live, ye shall live also," are the words of the greatest educator of man. Consequently the peace, which he left us as he departed, and which the world does not give, must finally become a fact. So also must every blessing which his life and his teaching would bring to us, that it to say, that union of life and Nature, humanity, and God which he anticipated and recognized. What he said about little children must be true—" Of such is the kingdom of heaven." His statement that " whoso shall receive one such little child in my name receiveth me" must be a fact, and not a glittering, specious but perishable sound of words tending to the indulgence of ambition and egotism, self-interest, etc., and even to the nourishment of sectarian hatred.

What has just been said in respect to the cube can be said with equal truth about the methods of guiding children, which have been already presented, as well as in respect to each of the modes of play and employment yet to be brought forward, as it is the keystone of the whole. For this is just the spirit that lives in the whole, in which and by which the whole has its existence, its being, its continued formation and cultivation. But, as it presses itself forward here in the midst as the center and middle point of the whole, who could or would hold it back? And so let it be here stated that this is now said once for all, but though said but once could and really should be repeated in respect to each exposition of it brought forward in the future, because it finds its more perfectly formed and more comprehensive confirmation in the greater increasingly cultivated manifoldness of the self-unfolding plays and ways of employment of children.

The law of connection is the most important law of the universe, of humanity, and of life in general. The child is to be treated as a member of humanity, and consequently of all life, in accordance with the highest and most effective laws of life, and is to be developed and educated according to those laws. But the child is also life itself. His plays and employments are mere representations of life; therefore the connection in his plays, and his ways and means of playing, must also appear as necessary as it is unsought. The next object of play is a proof of this.

The sphere and cube are pure opposites. They stand to each other in the relation of unity and plurality, but especially of movement and rest, of round and straight. The law of connection demands for these two opposite yet like bodies and objects of play a connecting one, which is the cylinder. It combines unity complete in itself in the round surface, and plurality in the two straight ones. The part of the surface of the cylinder taken for the base shows how in the first (i. e., in a curved surface taken for the base) is expressed movement, in the second (i. e., in one of the plane surfaces taken for the base) rest, as the cylinder combines plane and round.

Thus, then, the cylinder is the child's fifth object of play, and child life, especially in the country, proves to us the truth and correctness of the selection. Only consider how the country children play with the cylindrical or round pieces of wood, the so-called clubs, especially with shorter, more disk-like pieces sawed off from them, which is directly according to nature, since in these the plane surfaces predominate.

We see thus with pleasure that in the choice of these three early and almost first playthings of the child we have on the one side quite strictly followed the requirement of the thought, of the idea, and, on the other hand, the free life of the child and the requirements of that life, and so have come to one and the same result. These means of play and the mode of playing have approved themselves in the life of the children in freer, more playful development through the experience of more than ten years, during which they have been used with the child and with whole circles of children. This being the case we can be assured that we have seized the great means of development for the first stage of childhood which we have here in view.

There is nothing further to be said about the particular and individual use of the cylinder; this use is determined by its form as well as by the use of the two last-named playmates of the child.

But another essential thing must be expressed. As opposites with their connection, the sphere, the cylinder, and cube (as was before the case with the twice three play balls) appear as a play-whole complete in itself, and, belonging together as they do in a kind of family, form the second play-gift. We will later return to their more extended use.

Yet these first objects of children's play point somewhat toward the more effective phenomena of social life and of the life of art. For instance, the cube, the cylinder, and the sphere as a connected trio point toward another trio in architecture—the column, composed of pedestal (cube), shaft (cylinder), and capital (sphere).

This manner of connecting opposites by combining them into a whole complete in itself seems to me the most essential thing in the columnar orders. The tripartite character (always combining opposites) pervades their whole composition. I consider that on this connection and on this tripartite character is based the fact (as resting on a high comprehensive law of formation) that the columnar orders maintain themselves in their purity as the foundation of the noble art of architecture throughout Europe as well as America. The reason of this fact is that they not only form a beautiful whole complete in itself, but, in that whole, imprint on the mind a clear, simple thought, an idea full of life. Thus the Gothic style of architecture finds its value in the fact that it shows simple imitation of Nature in pure laws of number, form, and size, for example, in the trees standing together in orderly arrangement, although growing naturally with boughs bending toward each other, forming crossed and pointed arches.

This fact has been so distinctly brought out here for the purpose of indicating at some future time the uniting single spirit which makes itself known in all that has grown out from formation and through formation, especially from the clear thoughts and ideas of human beings at all times and in all zones. Another reason for the presentation of this fact is in order to point out the single spirit which at the same time expresses itself in human life and the life of Nature according to similar laws, although in different stages of development.

This spirit of humanity and of man, this spirit in itself single and therefore again leading to union in Nature and humanity (thus the single spirit of God working in all, creating all life, and again unifying it in higher consciousness), this spirit of unification must, like a warm breath of spring, spread over and unify the life of the child if we human beings are to unite and will unite in truth, first individually, then (socially and as individuals) in the family, in the life of the community, and in that of the nation; and if we are first to feel and then to recognize ourselves as a unit, in and with humanity, and finally are consciously thus to live.

This spirit of union must early light and warm the life of man, like the sun rising clear in the morning, shining anew over all life, separating indeed life and the living into their component parts, but yet again uniting all by the light which is in itself single. And " the spirit which forms life is unity." "One spirit it is which gives life unto all." So also the genuine spirit of the early nurture of childhood which awakens genuine life in and around us, gives it back to us and us to it. Only through this spirit is the long anticipated, long yearned for goal of all-sided life union gained by the individual man, the peoples, the nations, and all humanity.

We have now come with our play, our means of employment and gifts (in respect to the use of which we must refer, as we have before said, to life, to their employment in genuine spiritual kindergartens, and in families where the children are fostered in this spirit, as well as to the articles upon this subject which have already appeared and are still appearing) to an essential division of our subject, of the nature of which we have first to form a clear idea before we can with security go on building upon it and (proceeding from it in conformity with the subject) continue our lifeful development.

Now, what is the nature of the means of play which have been heretofore given? This, that the object of play was always a unit complete within itself, a non-separable whole. Only at last, at the conclusion, we see that three objects in themselves single can form together again a whole which is in itself single, like the ball, but in contrast with this a whole which consists of single parts bearing the same relation to one another and to the whole as the individual members bear to the whole which is in itself single; but, since they were considered in our earlier treatment always as wholes, they appear as part-wholes. And this property of things to be whole and parts, a part-whole, and the comprehension and treatment of things and so again of the human being, even of the smallest child in its earliest appearance conformably to this property, are so highly important for the child that the human being, and consequently the child, can not too early be led into the observation, knowledge, and treatment of the same. That this now is to be done by the next plaything and means of play is definitely expressed in the preceding (the second) gift; sphere, cylinder, and cube considered as a whole, in itself single.

But the instinct of the human mother also leads to this, since the mother seeks to content her restless child by procuring a plurality of things for its use. So also the child is instinctively moved to obtain a plurality of bodies capable of being again joined together, since it likes to divide separable bodies, and even likes to observe the parts of membered objects singly and disjoined and so as movable.

In accordance with this indwelling desire of the child for a divided and membered whole and the requirement for the next play-gift, already expressed at the conclusion of the second play-gift (as such a whole formed from different kinds of parts), now follows, according to the law of continued progress of the means of play, the cube once divided, but on all sides, thus again uniting in itself the opposites of one and all. By the division the whole cube appears divided into eight equal cubes, and we thus say, as it were, "see, it is highly important to consider the cube-form in life, therefore there appear here eight equal part-cubes."

And so it actually is. The knowledge of the cube form is so important for the whole life in all respects—in respect to the inner as well as to the outer life, in respect to Nature as well as to human life, and here again in respect to artistic, scientific, and practical life—that its form, its comprehension, and its management can not be too early or too urgently brought before the child. Demonstrations of this in life and proofs from life are innumerable; only two are here given. Winkelmann tells us that the Graces were at first honored in an old temple in Greece in the form of three perfect cubes. The importance of the cube for sculpture and for the higher and common architecture can be plainly seen. In science the knowledge of the cube form is, for the investigation of solids, what the straight line is for insight into the nature of the surface form; indeed, it appears full of meaning and significance even for moral life and its strivings. How instructive it is, therefore, in and for such an old and far extended association for general attainment of genuine humanity as a many-sided symbol of this effort and of the conditions for attaining its aim, etc. Therefore in the first and simplest division of the cube which the child at first attempts there appears only the cube again, though in numbers, and, as we significantly say from a higher insight into language, eight (acht) times presented to notice (Be-acht-ung). Therefore the attraction which these eight simple cubes, each of which appears exactly the same as the others, have for the children even in their second year, is quite magical. It is wonderful how these simple cubes are so warmly loved and valued in their constantly unalterable form and figure, as well as in their constantly abiding number, by the children who are ever striving for something new and different, and how fond they become of this small number of playmates which are always alike! (As the legendary world is of its dwarfs because they accommodate themselves to everything and are skilled and helpful to do everything, like the little mandrake [our Brownie]). Experience gives many proofs of this inward, delicate, cordial companionship, of these eight inflexible, unalterable cubes with the most thoughtful, delicate, actually angelic children in the genuine institutions for the fostering of children, as well as in the families where their most free and careful development is striven for, so that a whole pamphlet of lovely children's stories might be written about it. A few lines shall serve as a guide to the true inner relation of this plaything to the children and to their innermost need of development and training, for those who are not swayed by selfish motives (to mention whose names would here be inadmissible) but are urged by deep, holy, solemn earnestness to test by mind, spirit and life, experience and investigation, all that a pure, loving disposition toward children, human beings and humanity reveals for their education, so that there may again be for our children a genuine childhood, that is, a life in which spirit, mind, and power of action, feeling, thinking, and handling first penetrate and strengthen, as being intimately united, before they are necessitated to enter into the outer life in the separation which is an unavoidable requisite for higher consciousness. For separation is permitted for the observing, thinking, and comparing intellect and the outwardly representing life, and is indeed required by it, but must by no means on that account be permitted to appear in the mind which is destined to constantly grasp and retain in its original, inner union that which is outwardly apparently separated by the thinking intellect, the reason, and the life.

To fufill this (the most difficult requirement in life) or to live in life, which is esteemed the greatest art, can now never be taught to the *man* in the most highly remarkable and important ways, or, indeed, learned by him, just because it depends on comprehending and retaining the original (that which was originally in itself a unit). It can therefore only be *early* fostered, strengthened, and developed in the human being—the child. It must therefore also be early observed, in fact with the commencement of the life of the child. Now both these actually take place in a way highly worthy of thought, and most conformably and satisfactorily, in the way of observing and guiding, developing and educating children which lies before us for contemplation. And what is further not less worthy of notice is that just the point in this whole method of comprehending and treating children which meets with the most hostility, just the means of play most vigorously assailed by its opponents—the sphere, the cylinder, and the cube, and also not less the eight simple cubes—fulfill this most difficult requirement perfectly, lead most satisfactorily to the practice of that greatest of arts, since they foster and develop child life in the most delicate, judicious, warm, pure, harmoniously united manner just by the outwardly inflexible and sharp separation they present to notice, as well as to treatment, as is now proved by the life of many hundreds of children already occupied with them. The most striking proof of the comprehensiveness and deep foundation of the idea of genuine human training here presented is that it elevates to the highest glory just that of which it was to show the nullity. So it is also with the reproach that the children when occupied with Froebel's plays in Froebel's

Kindergarten are earnest but with faces not precisely either joyous or laughing. Indeed, life, with its deep earnestness, can not be early enough grasped, especially in the present times; and does not one of our most celebrated educators, Jean Paul, say, " Play has for the child the greatest earnestness and attracts the little one like a business; indeed, it lays claim to it like a kind of work by thought, feeling and acting, mind, intellect, and action." Thus here also—and it happens for the second time—the arrow which struck and was to annihilate the thing, fell as a trophy of war, feebly and without effect, before the feet of the thing at which it was aimed.

Rejoice, children, over Froebel's plays. They insure to you the highest treasure of life, a life which is in itself a unit. They insure you a thoughtful mind. They insure the anticipation of your hearts, and so confirm your blessed faith in an eternally invisible, single, good Being—God.

VI.

THE FATHER'S CRADLE SONG.

  Softly!
Softly, my son now sleeps, softly!
My boy's trust in me is deep
That he, through this stress of living
Be not roused too soon from sleep;
For the strength the night is giving
Through the whole day he must keep.
  Softly, softly, softly.

  Softly!
Softly, my son now sleeps, softly!
Childhood is of life the night,
Should in quiet stillness pass on;
Then, when comes the dawning light
Ready is the arm for action,
Strong and able for the fight.
  Softly, softly, softly.

  Softly!
Softly, my son now sleeps, softly!
In life's battlefield so stern,
Courage pure and strength uniting
Backward will all evil turn,

Gain the prize by bravely fighting;
This the weaklings never learn.
    Softly, softly, softly!

    Softly!
Softly, my son now sleeps, softly!
From the clear, soft light of dawn
Joyously the day is springing;
So the spirit's life is born,
To its deed its will thus bringing;
Thought of failure wakes its scorn.
    Softly, softly, softly!

    Softly!
Softly, my son now sleeps, softly!
Each act well weighed by the mind,
Prompted by the heart's warm glow,
Will fulfillment surely find;
Bravely thus be met also
Death itself—all ills combined.
    Softly, softly, softly!

    Softly!
Softly, my son now sleeps, softly!
So sleep on, my little lad,
Till to the heart's love warm within
You the spirit's light shall add
For strength and light the victory win.
Without light all life is sad.
    Softly, softly, softly!

VII.

THE CHILDREN'S GARDENS IN THE KINDERGARTEN.

The high importance of intimate acquaintance and union with Nature for the development of the child, the education of man, the training of the nations and of all humanity, has been already many times mentioned in this work, as it is the sure foundation of successful and profitable education and training of the individual as well as of all humanity. For we can not comprehend Nature in its whole being more precisely and more satisfactorily and vividly in all its relations than when we consider it as the direct manifestation in action of God, the first manifestation of God; but we have not yet developed the importance of this in detail. This importance is shown in the consideration of the growth and development of Nature, and its comparison with the growth and development of the human being, and so, first of all, considered and compared with one's own growth and development. If now this comparative study is important for man, it is especially important for the embryo man—the child. Thus an all-sided satisfactory education (and thus the existence of the kindergarten) necessarily demands that the child be afforded opportunity for this comparison—the word kindergarten tells us how and by what means, if we reflect upon its application—*in the garden of the children.* The kindergarten, the completely formed idea, the clearly demonstrated conception of a kindergarten, thus necessarily requires a garden, and in this, necessarily, gardens for the children. Yet the necessity of the requirement to connect a garden of the children with the kindergarten proceeds not only from the higher reason just given, but also from reasons of social and citizen collective life. The human being, the child, as a part of humanity must even early not only be recognized and treated as individual and single, thus as a member of a greater collective life, but must recognize itself as such and prove itself to be such by its action. But this reciprocal activity between one and a few, a part and a whole, is nowhere more beautifully, vividly, and definitely expressed than in the associated cultivation of plants, the common care of a garden, in which the relation of the general to the particular is clearly shown. This is a so-called house garden, but one in which each child has its place in its own little garden. But here in the children's garden of the kindergarten, where there are many children, and where they and their gardens form the principal fact, the arrangement must be somewhat altered. Here the gardens and respective beds of the children must be surrounded by the garden of the whole, as the particular always rests protected in the general, and the general protectingly surrounds the particular.

But this garden of the children, besides its general aim of representing the relation of the particular to the general, of the part to the whole, of the child to the family, of the citizen to the community, has to be in essentials not merely developing, educating, and instructive about relations, but also about things, and here especially about plants. But this is effected for the child in that he is summoned to comparison, and is again shown by the fact that the objects, here the plants, stand by one another for comparison.

In accordance with these ideas the following is submitted and stated for the arrangement of the ground, or the place for the children's garden in the kindergarten:

I. The total space of the children's garden has the form of a rectangle as the most suitable one. Other simple forms, circles and ovals, are not excluded, although they do not seem to correspond to the object of the whole, especially with many children, so well as the rectangle.

II. This whole space must now be divided into two parts—into the part for the general and into the part for the particular (that is, for the children), or, in other words, into the part for the whole and that for the individual members (that is, again, for the children).

III. The part for the general is the inclosing, as it were, the protecting part; that for the children, the inclosed, protected part.

IV. The children can not be and should by no means be introduced by this garden into the totality of the vegetable world, but only into the part which most closely touches human needs, thus into the field plants and those of the garden in a more limited sense; and therefore the general land should be divided into garden and field land.

V. But the garden land divides again into the flower and vegetable garden.

VI. The arable land is for the oil plants, corn, leguminous plants, bulbous plants, turnips and cabbages, finally plants for fodder.

VII. According to the quantity of land, larger pieces may be given to the children for their little gardens, and even a piece of garden land to each child alone. But if the children are many and the space limited, then the space for each individual must be circumscribed, or even a smaller piece may be given to two children together. This connection of twos in the kindergarten has something good in it—it teaches friendliness, and each child is so much the richer for what the other puts in the bed.

Where there is enough land each child may be

given four square feet in the form of a square; where there is less land, six square feet, in the form of an oblong, may be given to two children together. Where, however, the number of the children is large and the land small, two children must be content with four square feet.

VIII. The paths which divide and again combine the whole are either the principal paths or cross paths (between the single beds); the latter may be one foot wide. But it is a good plan to make the principal paths, if possible, at least two and one half feet wide, so that two children may walk in them side by side.

This must suffice for the division and use of the land in the general. In the particular the following is yet to be remarked:

In their own little beds the children can plant what and how they will, also deal with the plants as they will, that they may learn from their own injudicious treatment that plants also can not grow well unless they are treated carefully according to laws. This will be shown to them by the plants in the common bed, which they must observe carefully, so that they may calmly notice them in their development from the seed through the germinating, growing, blossoming, and fructifying to the seed again.

In sowing or planting this common bed the different seeds of the plants should be shown to the children for comparison and placed by one another, and the points of resemblance and difference should be sought out, so that the child may be able to name the different plants and to distinguish their seeds from one another. In the summer and fall, after the seeds have ripened, they should be gathered and kept for use in winter (about which more will be said in a later article on winter employments in the kindergarten) and for replanting in spring. When it is possible the seeds should be kept in little paper boxes which have been previously made by the children themselves. (More will be said of this in the articles about the employments of children in general and in particular.)

The plants set out in the gardens should be compared in the same way. Each child should have the care of keeping its own bed in order; the common beds should be cared for by all together, or by several children at the same time on alternate days (for instance, on Wednesday and Saturday afternoons).

In order that the children may be aware of the name of the plant while looking at and examining it, it is a good plan to place by the side of each kind of plant the name (formed by sticks) which can be easily read by children whose plays and employments have made them conversant with the stick forms; besides, the children thus obtain a clear view of and complete insight into the whole.

Just so manifold are the results when the bed of each child is indicated by its name in the manner before given. Each child immediately finds out the bed of his friend and himself receives, through the name standing by his bed, the merited silent praise or blame according as he has been careless or careful.

Besides, the child as yet unskilled in the knowledge of letters and of reading is thereby exercised in both, since he tries to find out the names from the signs.

Finally, and lastly, the child receives through all, as was already mentioned above in a more limited sense, a complete view of and clear insight into the whole, by which the memory especially is strengthened, the memory of places, things, names, and qualities, as well as the memory of time, by the differing stages of development of the plants, and by the tending of them.

All this, however, by no means exhausts the significance and influence of the garden on the children. As the child finds in it an image of the true family life, of the genuine common life (where the whole and general protect the individual and particular, and the latter has a retroactive, beneficial effect on the former), so he finds in each object by its creation, growth, and decay—that is, its development from the unit and its return to the representation of the unit—an image or type of himself which leads him to a better understanding, a more correct comprehension of himself.

It is an incalculable benefit for man to become early familiar with the course and the stages of his development, as natural as they are general (even if it be only in dim anticipation); and the boy or girl can early attain to this anticipation under suitable guidance by tending the little garden, and by observing the tending of plants by intelligent and experienced people.

What has been previously said will serve to explain the accompanying drawing (see page 226) so far as it concerns the representation and execution.

The garden is, according to circumstances, calculated for twelve or twenty-four children. According as the number of beds is increased in the length or in the breadth of the garden, the necessary number of beds for each number of the children can be obtained. Here a bed of four square feet is allowed for one child (or two children)—each cross path is to be one foot wide—each of the principal paths to be two feet wide. The width of the common bed which surrounds the small beds is likewise two feet, the length of this bed being divided into equal parts, according to the number of the fruits and plants to which it is to be devoted; here in the plan one foot (long or running measure), and therefore two square feet are allowed for each kind of plant.

The side A is devoted to the field fruits; the side B to the garden plants. The sequence of the former (A) and the manner in which the plants are grouped for the purpose of comparison are given. The sequence and the order in which the garden plants (B) are placed side by side for comparison easily result from the former. Since it is only too difficult, alas, to find with the kindergarten a larger place for the gardens or beds for the children, in a symmetrically arranged whole, the measures, especially the width of the paths, are here given as small as possible, and so the whole garden is only twenty-five feet long and fourteen wide. It is, however, better to make the principal paths at least two and one half feet wide.

Each of the sticks used for the child's name is given about one inch and a half for its usual width and about the same in length (which, however, may be three inches longer, according to circumstances). In the same manner one of the sticks used for the name of a plant is given in equal width, though not so long. The thickness of each stick is one fourth of an inch.

There must be as many of these sticks as there are of children and of kinds of plants. If it is desired to make the child familiar with bushes also, it can be done by a hedge surrounding the whole.

In a kindergarten which has lasted some years the seeds of natural (wild) plants (grasses, herbs,

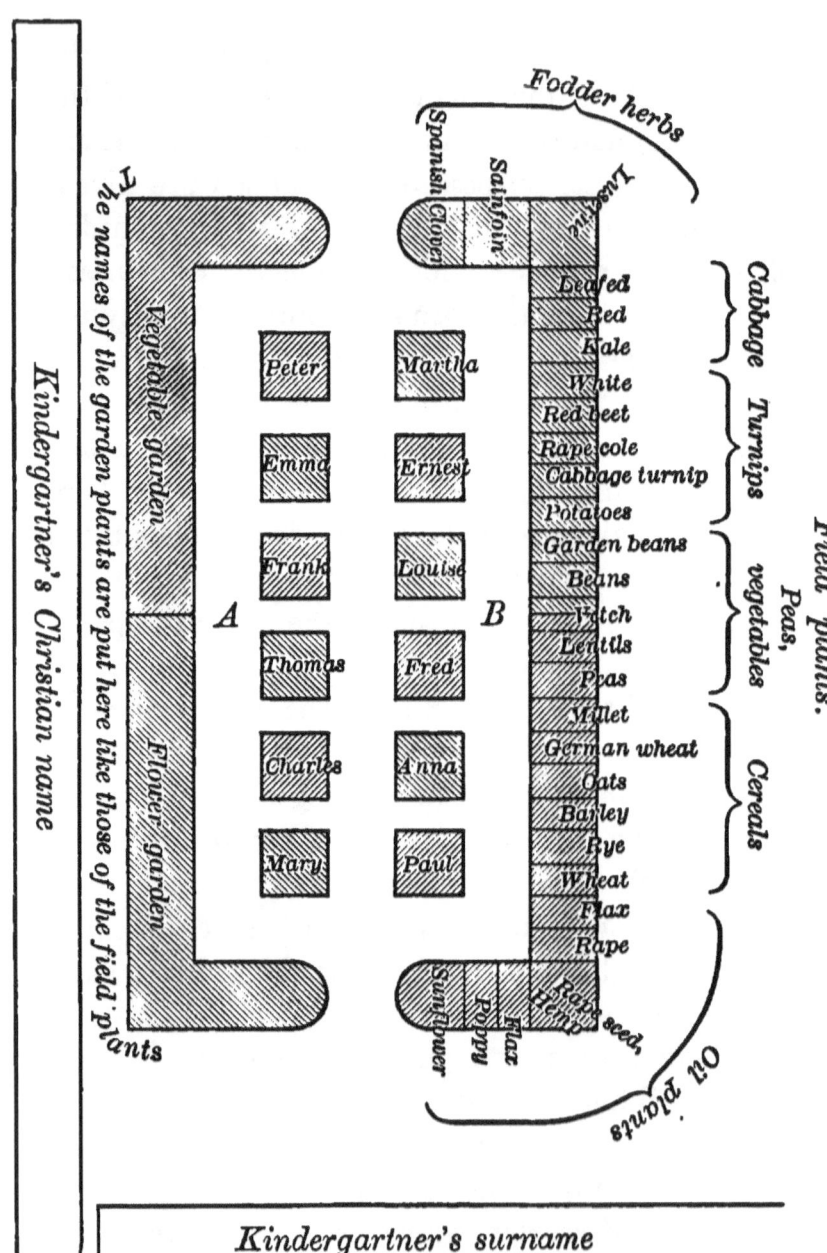

etc.) may be sown in the third or fourth year, instead of those of cultivated ones, thus increasing the child's knowledge of Nature and plants.

As the seeds and plants have been compared by the children in common, the seeds should be sown and the plants set out in common and, as it were, festively. In order yet more to fix this impression and expression, the kindergartner accompanies the planting of the garden with an appropriate song:

> Let us to the garden go,
> All our little seeds to sow;
> Warm air through the vale will blow,
> Making each seed sprout and grow,
> Etc.

So, later, when the seeds have germinated, the plants have grown,

> In the garden we would be,
> All our little plants to see.

It is only necessary to say a word more about the retroactive effect and the influence of such fostering of Nature and plants on the intellect and knowledge, as well as on the spirit and on the feeling, indeed on the whole actual and creating life of the child. For whoever thus stands in the midst of the whole, and thus grows forth in and from it, must well comprehend the whole. Therefore parents who possess a garden should never neglect to give up a sufficient space to their child or children to tend and cultivate in little beds.

They will thus provide for the children, with some plain, judicious guidance, a fount of inner moral elevation and strengthening.

Indeed even the thoughtful tending of a little window garden or a flower pot is for the child a pure fount of moral improvement. So cultivating is Nature in her effects, even through the simplest plant, for him who early opens heart and sense to her beneficent influences.

VIII.

PLAN FOR THE TRAINING SCHOOL FOR CHILDREN'S NURSES AND EDUCATORS.

"KINDERGARTENS" are the surest means, the most correct way, the simplest method of general elevation and ennobling, clear accomplishment and beautiful representation of genuine family life in all conditions and relations, as the single, true fount of contented individual life, joyful social life, free public life, and united life of humanity.

FR. FROEBEL.

§ 1. AIM OF THE INSTITUTION.

a. *In General.*

The aim of the institution is, in general, to train young women who are suited for such work, to tend, develop, and educate the child from its birth up to the time when it is fully prepared to begin its school life, and so up to the beginning of the instruction of the school properly so called. Its design is thus to train these young women in the union of the home, Nature, the school, and life as a whole (necessarily single in itself) for the attainment of the educational aim—comprehension, development and cultivation of man (from childhood up) in his personal individuality, and also as a member of the great life-whole.

§ 2. Aim of the Institution.

b. *In Particular.*

The aim of education designated in the former section can be attained in various ways, as it must be continued through various stages of development. These various ways can be essentially twofold, either the domestic family education (complete in itself) or the education in common in child circles and child unions (as one trains plants alone and singly in a room or in full life union with other plants in a garden), therefore the aim of this educational institution is also necessarily a manifold one, either:

I. The training of women as educational helpers for the house and family merely; and here again either:

1. First of all, only for the first stage of the fostering of childhood, the training of nursery maids and nurses; or

2. Up to the completely developed capacity for school, indeed up to the beginning of real school instruction, the training of directors and educators of children for the family; or, finally:

II. The training of directors and educators of

whole circles of children and child unions, as it were, true kindergartens (as one has flower gardens and tree gardens in which, as here, the flowers and trees, so there the children are the exclusive objects of consideration and fostering, of common development and education in the coherence of Nature and life), thus for kindergartners.

§ 3. Forming Plays for the Designated Aim.

The stated circles of educational operations jointly rest on the same foundation and are derived from the same source. Each one includes the knowledge and training of the others, only with greater or less conspicuousness and expansion of its own peculiar requirements. So the training of nursery maids and nurses (or rather, as they should be called, child fosterers and educational helpers) differs from that of the child directors and child educators, properly so-called, merely in this, that the training of the first aims more at mere practice and knowledge of the particulars and their true application, while the training of the second has in view more the insight into and the survey over the whole, and not only the freer appropriation resulting from these, but also the later fulfillment of the vocation, more full of life, and more freely active. Therefore the training for both aims of cultivation can go on in this institution in certain respects side by side.

§ 4. Age of Those who Enter.

A full-grown, healthy, and vigorous body fits those of from upward of fifteen to twenty years to be trained for the more limited, first vocation of child fosterers.

For the training for the more extended second vocation, as child directors and kindergartners, in general as child educators, the most suitable age is from seventeen to twenty years, in proportion to the degree of love of and kindness toward children, love and capacity for playful employment with children, and for the animated, joyous, and peaceful view of the world which they have attained. Yet, under the above-named conditions (which are indeed at times fulfilled in yet earlier years) older persons are not excluded from entrance into the training institution.

§ 5. Stage of Cultivation of Those who Enter.

Besides the already named conditions for the choice of this vocation—love for children, capacity and disposition for play and employment with them, purity of character, consequently sense and modesty—a womanly religious feeling of union with God, and a liking and capacity for singing, are indispensably requisite. The knowledge and dexterity which a good public school and girls' school give are also needed, especially by those who wish to cultivate themselves for child directors and kindergartners—in fine, as educators. Should the knowledge be more extended, so much the greater is its use for the future efficiency—also more extended—of those who enter.

§ 6. Duration of the Training Course.

Time and need press urgently, and the pecuniary means are usually insufficient to defray the cost of a longer training. So *fully twenty-six weeks* are fixed upon for the first course of training. On account of the inflexible demand of circumstances which I have just mentioned, this course must usually suffice for the acquirement of the most necessary training, as it can not be pursued further. Of course much might be compressed into the twenty-six weeks, and unremitting diligence, strenuous employment of the time, and some favoring preparation of the subject beforehand are needed to reach the goal. It has, however, been completely reached by several students.

§ 7. The Attainment of the Aim of Cultivation.

The attainment of the aim of cultivation indicated in § 3 (beyond the wise employment of the time required in the former section) depends especially on the students' correct comprehension of the impulses (to activity and employment) of their *own* lives, and of their laws, as well as, later, on the consideration and fostering of such impulses and of their laws of development *in the child*. The inevitable condition of the attainment of the aim of cultivation designated in § 3 is to see into these laws of cultivation, to find them in one's *own* life, as well as to recognize and foster them in the *life of the child*, or at least to faithfully order one's life according to their requirements, as they make themselves known in each simple womanly mind.

§ 8. Division of Time during the Training Course.

The day of instruction begins at seven o'clock A. M. during the winter half year for the participants in this training school, as well as for those in the general educational institution. At this time they take part in the general morning prayer and also, immediately afterward, in the religious instruction in the classes on that subject in the educational institution. They do this in order to obtain firm religious opinions and clear insight into the nature of religion and its development in mankind, especially at the stage of childhood. They should gain such insight for their own benefit as well as for later use with those confided to their fostering care, as religion is the only sure, satisfying, vital foundation of an education rich in results and blessings.

From 8 to 9 o'clock.—Breakfast and free time.

From 9 to 10 o'clock.—Bringing forward and observing the phenomena and course of development, consequently also the laws of development of the human being and the child; reflecting upon the insight into the being and nature of the child proceeding from these, and the demand thereby expressed for its tending and education, as a guiding thread to the fulfillment of the above-named vocation—the tending and education of children.

From 10 to 12 o'clock.—Acquirement of the before recognized means of setting the child to work and developing it, especially the acquirement of the right kind of intercourse, the suggestive talk with the child, learning the suggestive child songs, and especially the means of corresponding training of the limbs and senses for the unfolding of the soul-life in the child as a whole in itself, and also as a part of the whole human life. The family book [Mother-Plays, etc.] of Friedrich Froebel serves as a foundation.

From 12 to 2 o'clock.—Dinner, freedom of action, free employment with the children of the family circle, and the reviewing by each student of what has been before learned.

From 2 to 4 o'clock.—Making the proper objects for play and employment and the proper means for the developing of children, by education, for future reasonable self-dependence. Then practice with these objects for free use in accordance with the various gifts of Friedrich Froebel's play-whole; with essential consideration of all which Nature and life offer—that is, observation of natural objects and of the phenomena of life in accordance with the various stages of the development of childhood.

From 4 to 5 o'clock.—Supper and free time.

From 5 to 6 o'clock.—Taking part in the playful occupations of the little ones in the united educating families and of the youngest pupils in the general educational institution.

From 6 to 7 o'clock.—Acquirement of peculiar little manual dexterities which constantly develop themselves from the above-named play-whole, each forming in itself relatively an individual whole, and all together forming a coherent whole for the children's representing employment (on the side of the present domestic use, and later of that of the social, as on the side of the formation and cultivation of the sense of beauty and also of the laws of thought and reason), and for the awakening and anticipation of the inner coherence of all manifoldness, the development of a harmonious multiplicity from a unity, and for the anticipation of the laws of development in Nature and life.

§ 9. A Few more Essential Peculiarities and the Keystone of the Training Course.

The division of the day for the duration of the training course has indeed been given in general in the preceding section, yet it should be understood that even the advancing season effects many alterations in it. Thus, for example, in the spring months of the training course the science of plants for the stage of the development of childhood is entered upon from 2 to 3 o'clock.

There also intervenes between the above-given hours of instruction the execution of free exercises of the body and the practice of the movement plays; as also, twice a week for at least an hour, such plays with the little ones of the educating circle, and also twice a week occupation and play with the children in the school of our pastoral village of Eichfeld. As thus the training for the fostering and education of children embraces all the requirements of the child in bodily (dietetical) and intellectual (pedagogical) respects, so does it also in observation of all the directions of the various inclinations of the child, quite especially for the elementary preparation for the school and for leading the child into that preparation.

This preparation for the foundation of the school instruction, and the reference to it, can, however, be only briefly indicated in the first course, owing to the limited time, although sufficiently for a further individual progressive cultivation of the students through their ability and their own judgment.

Where a *second* complete course is possible, an elementary preparation for the school and the introduction into the first educational instruction is exhaustively given.

In this training course for harmonious, total fostering of childhood, as already indicated in sections 1 and 8, the child will be considered and treated in its individual personality, and also as an essential member of the great life-whole; thus, according to its nature, as a part-whole. Therefore one of the principal objects of the training school is to have the students learn to clearly know and vividly recognize their future little charges as they now recognize themselves in this double nature, as wholes and yet members, in order thus to educate them to the anticipation of the belonging together of all that is different, the union of all that is separate, to the anticipation of the invisible and spiritual in the visible and corporeal; of the abiding in the transitory; of the good even in the evil; and in this way to train them to the consideration and fostering of all good and to the union with all good in life, to the union in thought, will, and deed with Him who is good and the fount of all good—with

God—thus in genuine self-union and life-union, up to true union with God.

§ 10. Outside Conditions of Entrance.

The participants in a training course can either have board and lodging in the general educational institution, or the latter in certain houses in the village by the week, in which case — dollars are to be paid weekly to the above-named educational institution. The students themselves provide for bed, washing, light, and fuel, as well as for the care of the room. Doing the washing and the daily care of the room afford practice which may be essential at a later time, but the washing can be cheaply done in the village.

Or the students can procure board and lodging from the inhabitants of the place, in which case the compensation amounts to about half, including light and fuel in the common sitting-room.

One half dollar should be paid weekly for each person for the whole instruction, so that, in the first case, the whole course will amount to — dollars per month; in the second case — dollars per month.

§ 11. Beginning of the Course.

It is best for the training course to begin on the first of December, so that part of the time may fall in the spring months.

§ 12. Concluding Remarks.

The aim of this institution is generalization of an education corresponding to human nature and to the nature of the child, of an education which satisfies the definite requirements of the stage of human cultivation which has been attained by effort, and especially the more general carrying out of this education in families. The attainment of this aim has till now presented many difficulties in regard to the unavoidable expenses of training the students. Therefore unions of the fathers of families and other persons devoted to benevolent and humane objects, who recognize the comprehension of childhood and the aim of training here presented as founded on their convictions, are invited on the ground and in accordance with the aims of their unions to make it one of their objects to obviate these difficulties in cases where, for example, the establishment of a kindergarten in the province of their union is concerned, and thus to promote the efficiency of this training school; since all that is virtuous and good can be attained only through an education true to Nature and human nature, thus an education worthy of humanity, consequently the genuine family education.

But, on the other hand, young women in the before-mentioned relations and of the above-mentioned age who must choose a suitable means of securing a respectable living, and also those who are hindered by their parents or by their position, are invited to prove whether they find in themselves capacity and inclination for the vocation (so rich in blessing) of the education of children which is here presented; whether they find in themselves firm will, perseverance, and capability of self-sacrifice to seek out the right way and confidently to train themselves toward the fulfillment of this vocation. .

§ 13. REFERENCE.

All those who wish to take part in such a training course for themselves, their daughters, their relations, and wards, or for women as objects of humane assistance, should therefore address letters to the undersigned.

FRIEDRICH FROEBEL,
The Training School for Children's Nurses and Educators,
KEILHAU, NEAR RUDOLSTADT, *October, 1847.*

IX.

ADDRESS DELIVERED IN THE PRESENCE OF HER MAJESTY THE QUEEN OF SAXONY, IN DRESDEN, AT THE TOWER PAVILION, JANUARY 7, 1839.

NOBLE AND HONORED HEARERS:

In Nature as in life, all stands in constant, inner coherence, in the highest vital coherence which leads to God, indeed even unites with God. I believe that this highly respected, highly honored assembly will willingly excuse me from adducing proofs of this, and so much the more willingly as these proofs can be easily produced from every point of view, in every direction, and in every form, and can be demonstrated by the facts of Nature and life, as well as by the utterances of the wisest men.

In Nature all is intuition and life. Every phenomenon has its sufficient foundation and its necessary consequence. Finally, Nature is the first manifestation of God; it is the manifestation of God by fact and deed. Therefore this coherence is not only deeply grounded and true, but it is also equally deeply grounded and equally true that to live unrestrained and undisturbed in this high coherence of Nature and life gives to each created thing in its degree the finest fruits of life; but it gives to man the highest goods of the soul—serenity of spirit, peace of heart, and joyousness of life.

Was not this expressed and is it not still hourly expressed to us by Him who governs life, before whose name bow the hearts and spirits of those who know Him?

Since we now see represented in Nature for us, actually and symbolically, as it were, those highest ideas of life, since we perceive in Nature the fruits which are the sound and clear and complete living expression of the innermost, even manifestations of the highest (as the holy bards of ancient times teach us), should not therefore the human being, the crown of creation, also strive to live in this high, all-prospering harmony of life which God himself so visibly manifests to us in his creation and by his creation? Should now we adults, we parents, we educators and teachers, in general we fosterers of childhood and of humanity in the child —should, in short, the conscious human being, to whom it is indeed for himself no longer wholly possible to be able to live undisturbed according to the high coherence of Nature and life—should we not at least strive not only to lead our children, even from the beginning of their existence, into Nature and life in accordance with this all-harmony and coherence, but to make it perceptible to and recognizable by them?

The earnestness of this life question to which, highly respected, highly honored hearers, I could not close my mind and spirit, to the solution of which, therefore, I could not refuse my life and strength, is the cause that I ventured at least to strive toward this solution.

The earnestness and high importance of this life question caused me to venture to obey the summons to lay my attempt to solve this problem before an assembly, in every respect so prominently distinguished, for their examination. I beg, therefore, for great indulgence and considerate judgment of the imperfection with which I am now able to demonstrate the whole in word and deed.

If now there is everywhere in Nature and life inwardly united coherence, the truth of which has been conceded by us (or at least assumed to be conceded), each individual must be in life at the same time a whole in himself and a part of a whole—he must be a part-whole. The expression easily explains itself, as there is nothing in the surrounding world, and especially in Nature, which does not justify this expression, which does not illustrate its truth.

The child himself is such a part-whole since he unites and connects father and mother, the human being and Nature, the human being and God; thus he mediates between all being and life, even in its opposites, and therefore it is easy to awaken and foster in the child the anticipation, and, finally, the consciousness of the unity of the individual life abiding in God.

But how and through what is the human being to rise, as a child, to the anticipation, comprehension, and living expression of this inwardly united coherence and harmony of life?

First of all, the human being, even from his first appearance on earth, from his first entrance into the family, should be not only considered and regarded, but also tended and treated in all-united life coherence with God, Nature, and humanity, and in all ways as a part-whole.

Every error in regard to this threefold inner comprehension of the child and human being, even though it be but slight, injures his clear unfolding and disturbs the completeness of his life formation. Therefore this triune conception of the child and of the life of childhood is our first problem. Then, secondly, the human being, even as a child, is to be led to the anticipation and conception, as well as to the self-demonstration of the inward uniting coherence, so that he is led to observe, conceive of, and treat each object of Nature and himself first of all as a part-whole of Nature.

Therefore the real starting point, the sufficing foundation of all development is to retain this perception of Nature and of the world in its innermost essence and with all its consequences. But since each self-developing being and life, therefore also the life of the child, expresses itself in and by activity, we have thus, noble and honored hearers, come to the point from which I would like to begin the education of the child, or from which, in my opinion, a satisfactory education of the human being begins; viz., from the correct comprehension and fostering of the child's life and of his impulse to creative activity in accordance with the development of Nature, of life, and of the world.

But all activity is threefold, or expresses itself in a threefold action—in development, reception, and the union of both, viz., comparison and formation. This threefold activity appears in all which surrounds the child, as, for example, in every plant as well as in the life of the child himself. Thus the life of the child also must be comprehended in this triplicity of its creative activity, and must be treated according to it.

The human being, and therefore the child, is indeed led to all this through Nature itself, since there is nothing in Nature which does not lead from every point to God as well as to man; there is nothing in it (Nature) which is not a part-whole, which does not show activity. An apple on a tree is an example of this fact. It is a whole, since a whole tree can be again produced from it; and is also a part of the tree and of Nature, since it receives its juices through the tree and can actually develop itself into a tree, only in coherence with Nature.

But Nature with its phenomena is at once too near to and too far from the human being, and there is therefore needed, especially for the child, a connecting third which, as it were, unites in itself the properties of each part-whole of Nature and the properties of the child as a part-whole of the All-life, and yet is neither of the two. This is the ball. For since the connecting third can be neither an object of Nature nor the child or human being himself, it must necessarily be, on the one hand, a product of the human mind (so that the human being may be assured that it bears his nature in itself), and, on the other hand, it must bear in itself the properties of Nature, that it may be the mediator for Nature and for each part-whole of Nature. The ball fulfills these conditions, since it appears as a self-centered whole, but also at the same time as striving toward a higher whole, toward the earth and its center.

It is also the ball which especially corresponds to the life of the child and his activity, in which the child finds the self-centered starting point and completely satisfactory expression for this activity. As the ball easily develops activity from itself, it also receives into itself the activity of the child and shows both what is received and what developed in an externally uniting phenomenon.

I could therefore also call the ball the representative for all which exists, and this designation or view can not only be most completely justified, but even bears within itself something very developing and full of life. And so the phenomenon of the ball, as connecting the human being and Nature (and first of all the child and Nature), has its foundation in a higher indispensable requirement, viz., that the human being, in order to understand Nature, must create it anew, as it were, in and from himself in a manner peculiar to himself. In this effort now the plays entered upon by me and here brought forward for examination have their further ground as means of introduction into Nature for recognition and as types for the human being, first of all for the child.

But everything that exists like the man and child combines and unites opposites in itself, indeed resolves them in itself—for example, rest and activity, power and material, etc.—so it is also with the ball even in its outward appearance, since the ball (*full* of material) immediately requires and demands the *hollow* ball in the surrounding air or the surrounding general space.

Through all this the ball not only serves as an introduction into the knowledge of the most general properties of natural objects, but also into the knowledge of Nature and of the essence of Nature, and leads as well to the knowledge of the human being in his outward appearance as to the insight into his inner nature and into the unity of that nature, thus leading to the observance of its requirements. Therefore the ball is a mirror of all, and of the life in all, of the general as of the particular, and of the human being who receives both into himself.

To have intimated this about the ball and its nature as the first plaything of the child must now suffice for me, since I, noble and honored hearers, have been obliged already longer than I desired to direct your attention to this object in order not only to justify the starting from the ball, but also to bring it out clearly. For if we are united about the first point and the first means of the fostering and development of the life of the child we shall go on, I hope, like the skipper on the open sea when he has found the power or force pointing toward a unity—the compass.

But in Nature the outer proceeds from the inner, the special and particular from the general and united, etc., by means of the opposite; the inflexible from the movable, the manifold and composite from the simple and whole. So also representing all this, the inflexible sphere and the composite cube come forth from the soft and simple ball. Therefore the sphere and cube are the next playthings, or rather the next suitable means of employment of the child. But in the sphere, although solid, becomes yet more precisely prominent the easy mobility of the ball, as in the cube the manifoldness resting in the ball, of which I shall only permit myself to mention the most striking phenomena. I must wholly set aside, for want of time, the especial consideration of the sphere in contrast with the ball, since the phenomena of the sphere are only the sharper and more perfected phenomena of the ball, on account of the greater weight and hardness of the sphere.

In the sphere, as in the ball, in whatever position they may be, are to be distinguished three principal directions always at right angles to each other, conditioned by the lower, upper, and middle points. These three principal directions are in many ways the key to recognition, comprehension, and representation of every form and figure, of every size and number, even of life in its intellectual phenomena. It must not, therefore, remain unnoticed that as we perceive in each object three activities—the developing, the receiving, and the comparing—so also the principal direction, even in the ball, is determined by three points, and the principal directions in the ball, in the sphere, and in the cube are again necessarily represented by three. The importance of this observation of a constantly undivided, coherent three will come up to us, from this point on, more and more full of life, and will faithfully conduct us to our goal, as that perception, deeply grounded in the most secret innermost nature of things, faithfully shows, even to the child, the way to fulfill many of the requirements which life makes.

In the cube each end of the three principal directions appears extended to a surface, since each of the principal directions comes forward externally four times as four edges, whereby the cube shows three times four parallel edges. And so six surfaces of the cube, each two of which are parallel to one another, appear as postulated in its interior as well as in the inside of the sphere and of the ball. We must here postpone a particular reference to the manner in which, on the contrary, the phenomenon of the eight corners of the cube is presupposed in the inside of it and also in the inside of the sphere and of the ball. It is only important to us at present to give further prominence to the fact that in the sphere and ball all directions show themselves as different only in position, but alike in size. In the cube, on the contrary, these directions (which in the sphere only differ in position) show also the difference in size. They first of all show themselves as directions from one surface to another (surface directions), as directions from one edge to another (edge directions), and as directions from one corner to another (corner directions).

If we now look at this thoughtfully and scrutinizingly we must recognize that the cube, first of all, makes its inner nature externally visible; and, second, which is yet more remarkable, leads to the perception and recognition of the fact that the outside of the sphere is only the representation (manifestation) of the inside. A special mention of the manner of making this perceptible must also be postponed. Only the great law of Nature and life thus made known—to make the internal external, to make the external internal, and to place both in uniting comparison—this greatest and first (and yet simplest and most extended) law of Nature, life, and education, must not remain unexpressed. It is the more important to express this law as it thus becomes clearly perceptible how the kinds of employment entered upon by these plays not only lead to a notice of the laws of Nature, but bring them near to the comprehension and penetration of the human being, even as a child, and so to their application in his individual life. For we ask, What is the highest aim of man's activity from the beginning to the end, or at least what should it be? Can we give any other answer than to make known the unity of all life, goodness itself, God, through all which he creates from himself; to recognize the unity of all life, goodness itself, God, in all which surrounds and therefore acts upon man, and thus above all in Nature; and comparing both to perceive the one divine life, the unity in the workings of God, and, lastly, goodness itself everywhere in the universe, in life, and in Nature?

To this now I would guide even the little child (constantly bearing in mind the expression " one ") in his first free action, in his play, unconsciously to himself, and indeed wordlessly, through the contemplation of object and action.

It has been already recognized that the ball is the means of introduction into that which is general—first of all into that which is general in Nature. It must now be stated that the cube is the means of perception of and introduction into that which is particular in Nature, and also in the formation of the life of man and of the child.

A slight indication must suffice for the cube on each of these two sides. On the first, the side of Nature, this is thus shown:

Where the directions of surface are equally formed we find all the fixed forms which belong to the spherical, cubical formation, to the tessular system in mineralogy, as in the noble metals. Where the edge directions appear to define it we find the two-and-two-sided formation of the fixed shapes—*e. g.*, in the feldspar; and where the corner directions determine the formation, the three-and-three-membered system enters, as in the quartz.

In this there may at the same time lie an indication of how the child fostering for which we strive is at the same time also the genuine representation of comprehensive and impressive knowledge of Nature, and that the child therefore most thankfully recognizes a fostering of childhood in accordance with Nature when this kind fostering is also bestowed upon the natural sciences, since the child feels his relation to Nature in her innermost strivings, and so enjoys that fostering, although not arbitrarily and accidentally.

In reference to life, the fact appears that in all changes of the phenomena, the fountain and union of them, as well as their outer and inner coherence, should be retained for recognition and exercise. For the penetrating, thorough consideration of Nature leads the whole human race, as well as the individual man, always to God, as the divine teacher and educator of humanity himself says; and therefore the deeply penetrating consideration of Nature in reference to man's heart and soul, mind and life, becomes a sacred duty to his disciples and followers. Finally, who does not know, indeed what truly cultivated person doubts at the present time, that the thoughtful consideration of Nature has or should have a like aim with genuine child fostering, the aim which is given to the human being with his life, to recognize the fountain of all life, and, as we live through it, to live in and with it.

The sphere and cube will thus be given to the

child, and therefore in general to the human being, as the next plaything, or as the following means of employment, that he may be introduced into the manifoldness of Nature, of his own life, and of all life; and also at the same time that he may comprehend the inner and outer coherence of the two, and the unity in all manifoldness of each by the changes of the phenomena, and retain it that he may not be wrecked therein.

But I must hasten to my goal, however willingly I would still extend further my observations about each of these subjects, since I esteem them important for the life-fostering of the child and for the attainment of his future inner and outer peace of life.

As the unity of the ball requires manifoldness, especially the manifoldness of the cube and also of the sphere, as I have pointed out, so now further indivisibility requires the divided, the single requires separation. Each child shows us this, for he tries to divide everything; at least he brings each thing to his mouth in order to bite into it, although he is as yet unable to bite off a piece of it. This, however, by no means sets aside the fact that this phenomenon may have and actually has other causes. It is enough that we all know the child's desire to take things to pieces, which, if not sufficiently noticed in bringing him up, becomes a disposition to destroy.

The next plaything for the child must therefore be a divided simple body. This body must be divided in the smallest and slightest way, and yet on all sides. Such is the cube once divided on each side, through the middle and parallel to its surfaces, thus separating the principal cube into eight part-cubes similar to it, and exactly like each other.

Permit me one glance at our words, which are so full of meaning. On account of their importance, special attention (Acht) must be given to the eight (acht) equal part-cubes (similar in form to the principal cube), because what the principal cube shows once the eight part-cubes together make known eight (acht) times.

The division of the cube once on each side, or, in other words, its division on all sides, necessitates the remark that the opposites, which are in this case one side and all sides, must always appear united, and the one-sided should no sooner be seized upon than it is again recognized necessarily as a member, as an indispensably essential development of all-sidedness.

But now, how and through what does the once divided cube correspond to the nature of the child and satisfy his impulse to activity?

For application in life, especially in educational life, it may be stated that the tendency of the intelligence, proceeding from the first contemplation, is to separate; as the tendency of the heart, proceeding from the same contemplation, is to unite. This, however, does not deny, but rather proceeds necessarily from the fact that they, especially in later life, exchange their rôles and effects, so that what was before separating becomes uniting, and what was before uniting becomes separating. But this is here no further carried out, yet should be at least touched upon in order to avoid unnecessary objection. As now the intelligence separates and the intellect unites, practical life demands the form, and this is furnished by the cube thus divided.

So now the third play, which legitimately develops before our eyes, gives the possibility, indeed the necessity, of comprehending the child and the human being, by means of it, in the triplicity of his nature as a feeling and experiencing, as a thinking and recognizing, as a creating and forming being.

And so the play-forms which can be represented by it, or the results of the tendency to activity fostered in the child, are either forms of knowledge, of truth, of thought (often the child also briefly names them in his play, learning forms), such as $\frac{2}{2}, \frac{4}{4}, \frac{8}{8}$; or forms of feeling, of beauty, of the heart (the child also well names them, picture-forms); or forms of use, of life (the child also well names them, object-forms). As this exposition corresponds to the nature of the human being, of the child, so it corresponds to the phenomena of Nature, where appears first the thing and the object in general, then the object in reference to unity and beauty (as, for example, in the leaf and blossom forms), finally according to the laws which govern the organization of parts (measure and number, etc.). It must yet be remarked about these representations that a whole—that is, a divided cube—always serves for each of them; a similar law is also expressed in Nature.

But I must hasten on, since the briefest representation of this subject alone would fill up the hour.

It can not escape notice that each successive development must be already founded in the previous one, as this is a law of progression and development in Nature as well as generally in life. This is also an essential law of these plays, as a means of development for children. So the three principal perpendicular * directions of the cube appear even in the sphere; here, however, as changing; there as abiding, etc.

These three principal directions, although appearing permanently fixed in the cube, show themselves to be of equal value—that is, each of the directions can be put in the place of the other.

---

* The word is here used in its geometrical sense.—Tr.

If now any one of these directions is given—for example, as the direction of height—the two others are necessarily determined as directions of length and breadth. But the constant all-sided development of the child, as well as the introduction into the many-sidedness and all-sidedness of life formation, requires that with all changes of position these three different directions should appear fixed, constantly different in themselves from one another, yet each always the same.

Now, how is this to be obtained? Simply by carrying the dividing cut in the third division through the middle of each column, parallel to the two side surfaces, instead of through the middle of each column parallel to the top surface. The necessary result of this is the building-stone form, which is important for the forming life, especially for architectural life. The whole principal cube which before separated into eight cubes is now in this way divided into eight building stones, also well named by the children building blocks.

We see here again, according to a clearly visible law of Nature, the formation become more and more precise, and at the same time more and more manifold.

The forms represented with them by the children's impulse to activity are again, as before, separated into forms of life, beauty, and knowledge. But each series of them is essentially extended by representations on and in a surface, as well as by outline representations, especially from the province of life and use.

The most remarkable, however, is the phenomenon which is brought forward by this single alteration of the parts (that each direction, before only different according to the position, is now also different according to the size). This is that each single form of beauty or picture-form which could be represented, at the most, three different times by the cube once divided on each side or into eight part-cubes, may be carried out more than a hundred—indeed several hundred times—by the cube divided into eight building blocks. These forms are always different, yet are produced according to a simple necessary law without any arbitrariness.

What an effect it must have, what an impression it must make, even on the simple intelligence of the thoughtful child and on his later developed childlike intellect, when he perceives that the human being, even he himself, by his own small and limited individual power, can form such innumerable things with such small means! What must an unlimited single power, like that of God, be able to accomplish and form with infinitely less means! And these eight blocks have indeed proceeded and can be brought forth even by a child's power from the first principal cube, which by itself shows already such a great number of alterations.

According to experience, it must be assumed with certainty that such perceptions can not remain in the childish mind, in which often the smallest impressions have the greatest results, without beneficial effect for his whole life. And the observations of this higher yet simple kind are as many as the representations possible—indeed more, since each representation admits of several different perceptions.

These observations are so much the more important for the child, as not only can the representations and perceptions be again called forth at any time, but also the child is introduced by means of them from himself and his play into the notice of Nature; and his eyes, too, are opened to the intuition and perceptions in Nature where these truths become formed and actually shine wherever he turns. So through these plays and this way of playing in the life of the child is not only introduced and prepared for, but actually obtained and accomplished, that for which we strive—namely, that the observation and contemplation of Nature may be the foundation of human education, especially of the fostering of child life.

In this reference must be here made prominent a perception important for life and a general law which shows itself in these playful representations, as well in Nature as in the life of the human being and even of the child, and obtrudes itself in these representations. This law is that each following development includes each preceding and earlier one, as is demonstrated, for example, by the perception and development of the three different principal directions at right angles to each other, which are indicated even in the sphere, but first appear permanently different in the so-called building block. This perception is as important for the life of the human being, and particularly for the development of the child, as the law already previously brought forward, that all that follows must go out from that which precedes. But the child is led by Nature mediated by play, to the recognition of this law which is so important for his life and which expresses itself on so many sides, and indeed everywhere in Nature. This is the reason that these plays appear important to us. For as the child is led in such a way to perceive all development and manifoldness in Nature as proceeding from a unity, so again is he led through all manifoldness of Nature to its first unity, to its resting in God, to its having proceeded from God. This intuition of the mind, this recognition of the spirit, and this perception of life afford the highest prize of life to the human being.

Now a few more words about the next following play and its development from the preceding, as well as about its nature, etc.

Because the advance goes from the undivided cube to that once divided on each side, and because the most natural advance is from one to two, so the next plaything in this series of the means of play is the cube twice divided on each side parallel to its sides. This increased division, however, does not yet effectuate the progress of development in the child, for only a greater number of parts would be thus obtained. Therefore there must be added to the plurality of parts the variety and difference in kind of those parts, in order to obtain an actual advance. This is done by the division into the part-cubes, but necessarily, according to the natural law of the constantly requiring opposite, by a division wholly different from the preceding ones, viz., according to the oblique or diagonal plane of a part-cube, and once as well as twice crossing it. Since now, by the dividing of the principal cube twice on each side, the cube separates into three times three times three or twenty-seven part-cubes, it is in the nature of the thing that each of three cubes should be once divided into two diagonal halves, thus into two right-angled triangular prisms, in each of which two sides are similar and one is different. [This remark applies also to the bounding lines of each prism.] Then again each of three other small cubes is divided by two diagonal planes cutting each other at right angles into four equal quarters, thus into four right-angled prisms,* each consisting of two similar sides and a dissimilar one.

Indeed the correct comprehension of the right, in form and figure as well as in life and representation, is above all and first of all important for the human being and the child. Therefore, since Nature as a model for man teaches the right so multifariously, and first of all in the upright position (the perpendicular attitude toward the upper surface of the earth, etc.), so precisely the former plays sought also to confirm this comprehension and this impression. But the oblique and inclined is important for the comprehension of Nature, of her attributes, and of her laws, as well as of all the laws of life (who does not think here of the inclinations or dippings of the magnet and of their significance, as well as of the directions in the human mind, which are also significantly called inclinations!), for which reason these plays and means of employment of childhood in their progressive development in accordance with childhood also strive to train the child to the comprehension of the oblique and its extent in Nature, in art, and in life.

This may serve as an indication of the significance of this play for fostering the impulse to activity in the child, and for introducing him into

---

* The bases of these prisms are isosceles triangles.—Tr.

Nature and into the laws of life in general. The division of their representations into forms of thought or knowledge, into forms of feeling or beauty, and into forms of life or use (object forms) came forth clearly and importantly even in the two former plays and means of employment, and it does so here yet more. Since now, further, the number of forms is naturally very much more in each of the three branches or divisions than with the earlier means of employment, there soon come forth from each of them new and important ramifications leading into Nature as well as into life. Besides, the forms of each kind also appear on that account so much the more complete, the more finished and formed they are. Therefore if they are architectural forms they receive roofs, doors, etc.; if they are articles of furniture they receive more exactness; if they are forms of beauty they receive the essential, new feature that now, besides the square, the equal-sided triangle, and indeed the round which appears as a fundamental element of the forms of beauty in the most manifold way, and both appear as an introduction into the plant, leaf, and flower forms of Nature. The forms of knowledge present for comparison, besides an already large manifoldness of the simple or whole forms, a yet greater number of part forms, and as a wholly new result, a multitude of combined forms of truth or knowledge. I will cite especially the

Pythagorean Proposition, or the truth of the united square contents of the square surfaces on the longest side of the right-angled triangle, being equal to the square surfaces on the legs of the inclosed right angle.

This play and employment box is especially important for the child on account of the richness of its forms, leading on many sides into Nature, and on account of the multitude of simple facts of thought, life, and Nature which proceed from them.

Opposite to it in the series of play with tablet-like parts stands the cube, whose long tablets [oblong prisms] are again divided into square tablets and square columns. The constructions are, on this account, to a great extent columnar.

Yet, since it would be impossible in one statement to complete the whole of the subject in the yet remaining manifoldness of its direction and in the many kinds of their development, and still less possible to bring it forward with at least a few necessary illustrations, I close here my explanations of these educating plays. It has been at least possible for me to lay before you for your searching examination a few of the most essential perceptions, laws, and facts on which rests the effort we have begun, or rather from which it proceeds. You have received at least a general idea of the progress of the whole which further unfolds itself according to fixed and necessary laws; which progresses onward from this point, goes down by degrees more and more into the practical life of the child and the directions and relations of life awaiting him; comprehends his whole life, his inner and outer (that is, his spiritual and corporeal), his present and future life requirements at their starting point; and in such a manner that the whole comes to a conclusion with the development in his mind, of the conception that "through God and in God, by whom all is which exists, I also live and abide in the manifoldness of the phenomena of my life." In respect to the application of these plays in the life and society of children I must refer to what I was permitted to accomplish by kind, confiding permission and gracious favor here in several circles of children. Though it would not be possible for me to describe in one hour, even in its first fundamental lines, in the single accomplishments required by it, an idea very dear to man, highly simple, but rich in its development—to make Nature in its eternal laws of unfolding and life, placed in it by God himself, the foundation of human and childhood education—I do not therefore doubt of a favorable and kind indulgence. But if it should become possible for me to bring forward this idea (so simple in itself, and yet certainly beneficial in its results) so clearly that one would convince himself of its truth and possible applicability, especially with children before the age of school duties, I trust I should be pardoned if I, in conclusion, should express openly the wish of the heart and spirit which has faithfully fostered within itself this idea until now—for almost fifty years. This wish is that there might be for the demonstrated idea such a practical realization, a place for its fostering and unfolding, where it could make itself known in the entire fulness of its beneficent effect on the mind of the child, and in the rich blessings which it brings to the life of the human being.

Should this wish be fulfilled by the noble and respected assembly here present, it would thus become at once possible for the idea to give practical evidence, primarily by the blessing which develops from it for the life of the child, of the gratitude which is due for the indulgent reception of its first incomplete presentation.

X.

THE CONNECTING SCHOOL.—A LETTER FROM FRIEDRICH FROEBEL TO ONE OF HIS DISCIPLES.

MARIENTHAL, *May 25, 1852.*

DEAR AND ESTEEMED EMMA:

You desire from me an appendix concerning the management of the connecting or preparatory school—the connection between the kindergarten and the school for actual learning. This will, of course, be difficult for me, since I can not add the perception of the actual objects to my written words. However, I will make the attempt. First of all let us try to fix somewhat the different stages of the development of the child. The first stage is that of childhood. This is again separated into two divisions; in the first of which the tending of childhood, especially with regard to the bodily invigoration and strengthening of the child, predominates; and in the second division the careful development and use of body, limbs, and senses. This baby stage connects the child pre-eminently with the arms and lap of his mother. With the first stage is linked the second stage, which is continuous with it; this is the stage of the family. By means of the space in the room which is free to the child he develops to a wholly independent and spontaneous use of his body, limbs, and senses, and especially to a more complete development of his capacity for speech, so that he is at least in a condition to communicate all his needs, and to comply with the simple requirements of life which especially refer to alterations of space and determinations of activity. These two stages form a whole, in a sense, opposite to the kindergarten. The second principal stage of the life of children is that of the kindergarten. At its entrance into the kindergarten the child enters into a manifold new relation of life which should be carefully and thoughtfully considered by the kindergartner. The little one enters, first of all, into relations with a number of companions, and with those companions as individual parts of a whole, but he is himself also a part of this whole, and, as he has gained or lost from the whole, he has also duties toward it. In this lies the human training of the kindergarten, which the kindergartner must make clear to the child's consciousness in order to carefully introduce him into this new relation, and to make this relation fruitful to him. Second, the child, when he comes into the kindergarten, comes to a plurality of objects which lead him to comparing perception, thus to comparing afterthought, to the training of the understanding, and so, through their appearance and their relations unconsidered and unanticipated, to manifold recognitions. These objects become also for the child not only objects of perception, etc., but also objects of the activity of the creative will, and thus means for the recognition of his creative power and the results of that power. They thus teach him through the thing and the deed to know, first, the things themselves; second, their relations to one another; third, their manner of origin and development; and fourth, their further effect. All this the kindergartner must bring very clearly to insight and perception before she carries her child on to the third stage, the connecting school. In the kindergarten the question is merely of perception, contemplation, action, correct designation by words, as well as correct indication of what is brought out by action; but not yet of recognition and knowledge separated from the object.

Object and knowledge, perception and word, are yet in many ways as much united as body and soul. This training stage of the kindergarten must yet be retained (held fast) by the kindergartner as a very sharply bounded one. The abstract pure knowledge, the abstract self-dependent thought, is first entered upon in the fourth stage—that of the connecting school. The name closely indicates its nature. The connecting school stands in the middle between the kindergarten and the school for learning or for conceptions. It combines both, as in a certain respect it shares the nature of both; that is, it passes from the perception of an object to the idea of that object.

The precise and clear result of the kindergarten (complete within itself) is therefore a sharply defined and clear apprehension and conception of the object, of its properties, its relations, its origin, its onward development, and its manifold connection with life. And all this is connected with the accurately describing word, first of all, by the forms and images called forth by the child's freely creating activity, as forms of life, as forms of knowledge (recognition) and insight, and as forms of feeling, forms of beauty. Here that which is inwardly single and existent appears in outward manifoldness, and so the child comes to the true recognition, to the particular perception of the manifoldness which resides in the inner unity and comes forth from it by legitimate unfolding. This is one of the most important phenomena to which you must call the attention of all those who examine the subject; for, as all proceeds from a unity and again returns to a unity through manifoldness, oppositeness, and connection, its contrary and opposite is given with each thing, and so the child is introduced unconsciously into the science, indeed into the living expression of the simple and general as well as the special laws of life by this definite action and by this holding fast in feeling and comprehension; the child indeed lives in these laws. You must, first, make this very clear, vivid, and intelligible to all; and, second, show it particularly to the examining authorities, and especially to the open or silent opposers of the system.

Therefore the stage of kindergarten training must be very clear to you. Your perception of it must be very decided. In this kindergarten training the law of development, constantly used and followed, counteracts the impulse to destruction, especially in boys, as it arouses the impulse to developing, creative, formative activity.

But what is the result of this developing activity? It is as follows: That which is not apparent becomes evident (in the sphere one recognizes the axis); the invisible becomes visible (this applies to the child). In the action of the parents the child recognizes their love, and *vice versa*. In the manifestations of Nature the oneness and the love of God are disclosed. To this disclosure of that which is not apparent lead the stick-laying, the interlacing, the intertwining, and especially the peaswork, rising from the hollow, empty, plane surface to the hollow, empty solids—cube, octahedron, tetrahedron, etc.—and their constant connection with and abiding in cube and sphere. You have been through all this work and can consequently demonstrate it also to your examiners and critics.

Before the stick-laying I directed the laying of surfaces or little tablets. The laying of squares and triangles especially should be mentioned. This laying of surfaces leads to the highly important knowledge of the relation of form to contents, or of figure and form to size. It teaches us to know the laws which lead us into the deeper knowledge of the nature of things—namely, that like form is possible with unlike size, and that equal size is possible with unlike form. The surface-laying also teaches us the laws of size and form (mathematics), which develop further from those just mentioned, and shows us how the *first* apprehension and perception of these laws is merely a simple making, changing, doing, without any further reflection and without any words.

You see, my dear Emma, you must expound the nature, means, and ways of kindergarten guidance in such an organic manner as this to your educational officials as well as to the severe critics. If you do so you are provided with weapons and will be unconquered, even if no one agrees with you, even if no one says that you are right. Your being right does not at all depend on the acknowledgment of the fact. You can be perfectly right without the acknowledgment being made by another, just because he does not see into the subject; insight can be impressed upon no one. The being right is based on mathematical proof which no one can oppose, and which speaks through the object.

But the whole kindergarten procedure rests upon simple mathematical proofs, and you yourself must rise to the perception of them. Now, there are two more principal perceptions of life which begin in the kindergarten and which are demonstrable in their fundamental generality. These are the relations of plurality, mass, number, to unity, and the relation of the designation to the thing; and here again in a double reference, first that of the word to the thing, then that of the sign to the thing.

As the numbers in their essential diversity as even and uneven numbers, as three times two and two times three [square surfaces], are seized in the cube as unity, this is an important fact of our kindergarten procedure. I place great value on this, as I do on everything in which manifoldness develops from unity through contrast and again returns to unity. However, number first finds its true recognition with the stick-laying, where, as you know, all which teaches the relations of enlarging quantity to increasing number, the distribution of the size of the parts (thirds being smaller than halves, sixths than fourths) comes out necessarily and in the simplest way. Thus the foundation is laid in the occupations of the children in the kindergarten for the perception of number and its relations, as so-called integers and fractional or divided numbers.

You must once more quietly recall this to your remembrance, and bring it to objective perception in your own room, dear Emma, so that you may again obtain the feeling of supervision and mastership. You should never allow this feeling to be weakened, my dear Emma, but you must, on the contrary, strengthen and elevate it.

Now let us turn to the consideration of the representation of the object by word and drawing. The former, in the kindergarten, is confined to the spelling and writing of a few names, and the article entitled How by Means of Persons and Things Lina learns to Read * gives you sufficient explanation of it. If you develop what is there stated in and from itself, as your own view, you can thereby meet each criticism and question with the full feeling of sufficiency.

This subject is treated in the just-mentioned article so lucidly, so truly, so in harmony with the development of the child and the reflective nature of man that it merely needed to be read to establish the truth of what is there stated.

Next comes the representation of the object by drawing—that is, by the sign. Little of this belongs in the kindergarten, because the little fingers

---

* See vol. xxx, International Education Series.—Tr.

THE CONNECTING SCHOOL. 277

are as yet too weak. Stick-laying represents the drawing in one aspect; the making of circles, which the children like so well to do with the slate pencil, is another aspect. The latter can be carried out to the simple flower and leaf forms.

However, on account of the weakness of the little fingers, the drawing, as well as the writing, belongs predominantly to the connecting school, as do the practice in color and the genuine singing exercises, for which the singing in kindergarten is only the preparation.

Add to this the introduction into life itself by the movement plays, by the tending of the little individual gardens and of the general garden of the children, and by the personal feeling of selfhood and life awakened and nourished in the child by the play and the tending. Add also the presentiment (caused by the just-mentioned feeling and aroused at the same time with its increase) of a fatherly Giver of life, and of the feeling of his fostering care of life, as the foundation of which may be claimed the testimony of Jesus, that children wish to be good. Put all this together and you have the kindergarten in its completed cultivation, and the child, as a member of it, at the threshold of the connecting school.

Here presses on us pre-eminently the question, Then what makes the connecting school a connecting school? The name says clearly that it makes the connection between the kindergarten and the school for genuine study, and is a passage from the one to the other. The name also implies that the connecting school comprises and unites within itself the nature of both, proceeding from the development and nature of the kindergarten to the school for genuine study, and to the right guidance of the child in a manner corresponding to and faithful to his nature and its requirements.

Now, what is the nature of the kindergarten? And what of the school? The nature of the two may be thus described: In the kindergarten the principal consideration is the child, his nature, and the strengthening, invigorating, developing, drawing out, and educating of the little one; in the school it is just the reverse. Here, in the connecting school, the principal consideration is the object, its nature, the recognition, perception, and comprehension of its properties and relations, and the designation of those properties and relations; the training of the child thus effected is but secondary, incidental, and casual. Through the demands on the child to recognize the object, the fact, the thing, in its right nature, in its true properties and clear relations, the child is still considered; but the correct comprehension and knowledge of the object through perception is ever the principal consideration. In the school the principal consideration is the comprehension of the object through thinking, the inner presentation, as it were the unclothing from the body—abstraction.

The connecting school thus forms the step from the perception of reality and facts in the kindergarten to the comprehension of abstractions and of thought in the school. You must make this very clear to yourself, my dear Emma, and you are right in thinking that the correct comprehension of the nature of the connecting school and its proper guidance is very difficult, or at least not easy, just because it presupposes exact knowledge of the kindergarten, its nature and peculiarities, and also at least the general knowledge of the school, its subjects of knowledge, its nature, and its demands.

Whoever, therefore, sets for herself the task of completely carrying on a connecting school must go through a training course of at least one year, if this guidance is actually to lay claim to completeness and perfection. On account of this lack of thorough training for the connecting school it is for the most part carried on so imperfectly, and on account of the twofold character of the training required, in spite of its high and great importance to the teacher, and even to the public teacher, it is still so rare.

Upon what path does the connecting school now enter? It connects accurately with the facts and phenomena, with the sense-perceptions in the kindergarten, but gives generality of significance to the observation of particulars, and thus gives intellectual comprehension and a form of thought. For example, my little ball moves easily, there and here, forward and back, up and down (kindergarten perception). Everywhere in space I can think of three lines, of three directions, all three of which intersect at one point, and at right angles to one another. I can draw them and place them before myself in thought (connecting-school conception). Or, going yet further back, "the little ball escapes from my hand and hops out free." It escapes from the small, narrow, inclosed space into the great free space (kindergarten perception). Every object can move in particular or general spaces (conception of the connecting school). Exercise in the connecting school: What rests or what moves in particular, and what in general space? To connect with the preceding kindergarten perception: How does it (state it in a general form) move in space? Answer in the connecting school: In three directions at right angles to one another (comprehensive form of thought). This form of thought explains in the connecting school the diversity in the movement of the ball. Or, again, one whole, two halves; two halves, one whole (kindergarten perception). I can divide each whole into two halves, and always unite the two halves of a whole again to form this whole (intellectual and general comprehension of the connecting

THE CONNECTING SCHOOL. 281

school). Where do whole things appear actually divided into two halves, or halved? and what wholes can I divide into two perfect equal halves? (progressive development of the connecting school at this stage, used for the perception of number and quantity). Passage to figures (distinguished from number). "You see, children, it would be too tiresome always to make the proper number of strokes for the numbers or quantities, so people have found out signs for the numbers or quantities (first of all up to nine) which are called figures. These figures may perhaps have originated in the following way:

$$1 = 1, == = Z = 2, \equiv = 3 = 3, \mathcal{L} = 4 = 4,$$
$$5 = 5 = 5, 6 = 6 = 6, 7 = 7 = 7,$$
$$8 = 8 = 8, 8\text{)} = 9 = 9,$$

But we have a particular sign for each particular number or quantity. In other words, we have a particular figure for each number. A number or quantity of | | | | | | | | | | is considered as one single composite (drawn together) whole, and is hence called the ten (zehen from ziehen, to draw). This number is again indicated by a figure 1, but this figure occupies the second place counting from right to left, etc." (Connection with arithmetic.)

It is remarkable how the whole of arithmetic, and the whole teaching of number, is connected with the perception forms of the kindergarten, and you, my dear Emma, can most completely satisfy each examiner of the subject, and each critic, of the truth of this statement. Alas! I can not here demonstrate it all to you just because it presupposes the direct perception and presentation of the object, which I have neither time nor space to supply here by written words.

But as the connecting school leads so deeply and fundamentally into the science of numbers and figures, into real calculation (arithmetic), so also does it lead into the knowledge and science of space, form, and size in a greater or less circuit—the circuit as much greater or less as suits the views of the teacher, the need of the scholar, or of the school in general.

Willingly as I have shown you here the easy and satisfactory way conjointly with the word and the perception, yet, as I must refer you to my Education of Man for my treatment of number, so I must now also call your attention in respect to form and size to the plates illustrating the Fifth Gift.* You may also recall to your mind the forms of knowledge with the different triangles. These and the peaswork are the most important means of passing from the kindergarten, through

---

* See vol. xxx, International Education Series.—Tr.

the connecting school, to the school for study, thought, and teaching. They also form the most important means of connecting the former with the latter, which, to use a common phrase, we will term "the school of instruction." I use these different expressions in order to give you an opportunity of grasping as exactly as possible the nature of the school, and so of comprehending the nature of the connecting school.

You are quite right. The keystone of kindergarten employment is the transformation of solid bodies, and consequently the knowledge of the relations of the different geometric (crystalline) solids to one another, as well as their development from one another, and the relation of all to the space-filling, geometric unities.

The fourteen solids which you have received, and, if possible, the making of such forms with potter's clay or with cubes of turnip or beet-root, furnish the means to acquire this knowledge. It is best to have the cubes of equal size prepared by the joiner, if he will utilize for this purpose his somewhat disused tools. If it is difficult for you, however, to have these prepared you must content yourself with the box of fourteen solids which was sent to you, and with their derivation and development from the cube.

Let me first recall to your memory the use of this box of fourteen solids in the kindergarten.

The training of the human being which educates by developing, or the kindergarten training which is a complete whole, begins, as I have already said, with the tending and observation of the stage of infancy, and here again with personal and extraneous care (the care of that which is foreign and external to self), and the comparison of both. The second stage is that of complete development of limbs and senses and of the body. Such development is needed that the child may do justice to his own personality, and may bring things near to himself, or himself near to the things in order to look at, to handle, to use, to utilize, to change them.

For this purpose the child must learn, above all, to know each thing in its capacity for filling space, in its property of being defined within itself, in its rest and motion, in its form and size, in its gravity, etc. The ball serves this purpose for the child. The ball unites in itself and shows all the properties which appertain in general to all and each object which has a body and occupies space. Since the ball shows boundary, visibility, and invisibility, it leads to the great law of the world—the law of opposites (contrasts) and their connection. It leads even to the discernment of that which is within, of that which is invisible and single (in the middle), as well as to the connection, that which is visibly invisible (in the axis). It also leads the child to the discernment of all the fundamental properties of all things and of each thing, and at the same time introduces him into the outside world.

According to the law of opposites the soft round ball demands the hard sphere which shows yet more precisely many of the properties of the ball.

The round sphere, having one surface, no edges, and no corners, requires its opposite—the cube, which has straight surfaces and more than one surface, and has also edges and corners—and leads thus to the manifoldness of the properties of things by holding fast their inner invisible unity. The sphere also now appears clearly as the expression of motion and of easy movàbility, and the cube as the expression of its peculiar gravity and rest.

The cylinder shows the connection between the two, the motion in a straight direction connecting rest and movability.

Now, you know the second play gift of the children—the sphere, cylinder, and cube, the children's delight—with the richness of its phenomena. With this gift is now connected the employment with the fourteen solids, which employment presupposes the former. Therefore sphere, cylinder, and cube—the latter in its twofold form, first as a mere mathematical cube, second as a cube prepared for manifold alterations by being pierced and having wires in it—are the first four solids.

If now the child considers these three (relatively four) different bodies in their different phenomena, what have they shown and taught him? The answer is given by the connecting cylinder.

The round would fain unite with it the straight, and the straight would fain unite with it the round. The cylinder results from this reciprocal effort, from the union of cube and sphere, as it were.

Therefore the points would fain become surfaces and lines, the surfaces would fain become points and lines.

Suffice it to say that each would fain form and develop (as it were, be the living expression of) all others.

We see, therefore, how the inner organic and living law of change, of development, results from the apparently outward law of contrast and connection. You will remember that this manifestation and this law already came forth in and with the employment with the right-angled isosceles triangles where the outward, mechanical, inorganic grouping led to a living, inner, organic coherence. What took place there with the surfaces, and with the interlacing and intertwining at the stage of lines, takes place here in predominantly increased completeness at the stage of corporeality, and of the capacity of bodies to fill space. Hence the contemplation of the fourteen solids introduces us and the child into the province of formation in Nature, and first of all into the province of formation of the solids.

The cube (or dice, from its original use) is also familiarly called the hexahedron. The octahedron and dodecahedron also receive their scientific names from the number of their surfaces.*

According to the law and effort above stated, the corners seek to extend themselves to surfaces till the surfaces touch one another and, as it were, reciprocally set limits to their further formation. Thus results the cube, with its angles replaced by planes.† Its place in the play-box is in the third compartment of the left middle row; the eight completing forms lie in the first box of the left completing row.

If you will have this done, dear Emma, first of all in the kindergarten by the children who are in their last quarter (in which it belongs) with potter's clay, or any other material which can be easily cut (turnips), you must give the children perfect cubes of this material and require them (by few or slight, or regular cuttings off of all eight corners) to change each corner or point into a plane or surface according to the desire

---

* The German names for the two latter solids indicate the number of corners and of edges respectively.—TR.

† Literally the six-eight surfaced (sechsachtflächner)—that is, a solid with fourteen sides, six of which are similar in form and size (octagons) and the other eight are equal triangles.—TR.

of the cube (as it were) till the surfaces touch in the middle of the edges of the original cube, and thus results from the activity of the children themselves the six-surface previously produced and named.*

Now, if you have sufficient material you can let this solid stand for itself in the degree in which it has been formed, and do the same with a new cube up to this point. Then continue by cutting off thin slices till eight equal three-and-three-sided † surfaces, which become on further slicing hexagonal surfaces; and, last of all, purely equilateral, triangular surfaces appear lying opposite ($\triangledown$) to the former ($\triangledown$). The six cube surfaces have wholly disappeared. In the place of each appears a four-edged corner; and in the place of each former three-edged corner appears now a purely equilateral triangular surface. In the place of the hexahedron (cube), and as if from within it, appears the octahedron, and as a connecting intermediate form appears the six-eight-surfaced solid which was given by the first stage of change.

In the box of the fourteen solids the mechanically organic development of the octahedron from the cube could, alas, be but very incompletely shown, since equal, whole corners would have to

---

* Froebel says sechsflächner here, referring to the "six-eight-surface" solid described above.—ED.

† I. e., three long and three short sides to each.—ED.

THE CONNECTING SCHOOL. 289

be taken away. However, by means of this box the child learns at least how and where the octahedron lies in the cube.

N. B.—The supplementary forms to the octahedron lie in the first box of the completing row, quite at the right hand, the opening of the box being turned toward the teacher.

We now go on, dear Emma.

The effort to become surfaces was shown and accomplished by the corners, so that the six-eight-surfaced solid and the octahedron resulted from the cube. This is also shown by the twelve edges of the cube. This effort can be illustrated by means of soft masses in the same way that the corners were changed, or by taking away the twelve completing forms. Suffice it to say, there results, in the double way before shown, from the cube

first the six-twelve-surfaced solid;

then the pure (rhombic) dodecahedron.

See the six-twelve-surfaced solid in the fourth compartment of the left middle row, its completing forms in the second box of the left completing row, the dodecahedron in the fourth compartment of the right middle row, and its completing forms in the second box of the right completing row.

Cube, octahedron, and dodecahedron, with their connecting forms, the six-eight surface and the six-twelve surface, are the chief forms and figures of the three equal surface directions of the cube, which intersect one another at right angles, which

lead to the sphere $\begin{cases} 1, \text{ by the surfaces;} \\ 2, \text{ by the corners;} \\ 3, \text{ by the edges.} \end{cases}$

Yet where powers and efforts move, there is also a heaving and a pressing forward and back, and, finally, a suppressing; so, first of all, with the surfaces of the corners, four corners are wholly suppressed, and four corner-surfaces come out prominently; thus are found

first, the six-four-surfaced solid,
next the pure tetrahedron.

See the former in the fifth compartment of the left middle row, and the latter in the fifth compartment of the right middle row toward the outside, and the completing forms of and to both to the left and right in the third box. We may also look upon this formation as if it arose from six angle-diagonals or oblique lines, touching at their ends, which form edges.

The development of these six new solids and the knowledge of their outward relation to one another can close the course in the kindergarten with the sixth year. The reverse course, which is almost easier than this, can, however, be also given, and the cube, the cube with its corners replaced by planes, and the octahedron may be formed from spheres of soft loam, clay, or sand, by equal cutting off of two, and two, and two opposite points, each two of which are opposite to each other and their connecting line of direction is at right angles to the two other connecting lines of direction.

Therefore it is more important at the stage of kindergarten employment and training to raise that which can not be seen through to that which can, and to raise that which is material to a less material perception, than to advance to the solids of the edge- and corner-diagonal or oblique lines. So you have already made, with your little ones, the square (each side as long as two of the edges of the cube); the rectangle (oblong), which is half the size of the square; the right-angled isosceles triangle, also one half the size of the square, divided by a corner- or angle-diagonal line, etc. This is clear to you, as you have already done it yourself so many times.

The outlined cube results from two squares made of sticks exactly equal in size, joined by four vertical sticks each of the same size as each stick of the squares.

The outlined cube with its corners replaced by planes (six-eight-surface) originates within the cube by making the opposite square (i. e., one-half size) in each of the surfaces of the outlined cube.

A pointed column or pyramid composed of four equilateral triangular surfaces erected on the two opposite sides of an opposite square gives you the outlined octahedron, as you well know.

Take six sticks, each of the length of the corner diagonal line of a principal square, connect each three ends of these lines by a pea (pieces of cork or wax), and, as you know, the outlined tetrahedron results.

Now you know that all these bodies with a common, invisible, middle point can be represented in one and the same cube in an easy way.

Yes, and with these representations, their comparisons with one another and their comparison with the collective bodies which can not be seen through—that is, the solids—with the perceiving and demonstrating of the one in the other, there is stated and included at this stage of active occupation and kindergarten employments an intimation of the development of manifoldness from unity, of the invisible from the visible, of the inner from the outer, and the reverse—that is, the development of perception from conception and thought, of thought from action, of conception from desire (will), etc. The child is ripe for entering, and is quite sufficiently developed to enter the connecting school. He stands on its threshold, before the door. The child steps into the connecting school. (The child of the kindergarten is now a little boy or a little girl.)

That which was the keystone of the kindergarten training is now the starting point of the first stage of the connecting school—the particular,

THE CONNECTING SCHOOL. 293

individual, and objective perceptions of thought. Therefore,

Opposite to unity is singleness.
" to singleness " manifoldness.
" to the outer " the inner.
" to the visible " the invisible.
" to the simple " the complex.
" to the round " the straight.
" to motion " rest.
" to the whole " the divided.
" to the single " the composite.
" to the quiet and abiding inner being is the outward appearance, the becoming.
" to the outward grouping is the inward coherence.
" to the outward combination is the inner development.
" to the passing appearance is the abiding effect.
" to the mere effect is the life.
" to life " the living.
" to the living " the sensible.
" to the sensible " the rational.
" to the unconscious " the conscious.

} And the reverse.

We can add as a connection: to unconsciousness, coming consciousness.

To coming consciousness, consciousness.

To the mute and yet speaking form, the clear, speaking uttered, audible word.*

You see, my dear Emma, that the child passes from the kindergarten into the connecting school, being skilled in and capable of quite precise perception and conception of all these contrasts. These

---

* The life and works of man are opposite to the life and works of Nature, etc.

21

contrasts can and should be brought to the child's observation as opportunity offers and necessity requires, but this should always be done in a sequence similar to one of those indicated. Now see with what a foundation, with what a groundwork, with what an amount of germs of life in the collected material of life the child passes from the kindergarten into the connecting school. At no point of life is the leading direction lacking. The germ for each development required by life is provided, as is shown in the great whole of Nature. All this awaits only the development from unconsciousness through coming consciousness to consciousness, and this is the task of the preparatory school; the keystone of the kindergarten is, as I have before stated, the first stage of the preparatory (connecting) school.

The nature and character of the connecting school are in many ways clearly indicated; the particular is advanced to generality, the outward isolated perception to an inner total conception—for example, one and the same child lets the ball jump from his closed hands into free space in different parts of the schoolroom, and he himself moves in three principal directions (at right angles to one another), there, here; forward, back; up, down; or several children do this in succession with a ball attached to a string, or several do it at the same time. Suffice it to say that in all these cases the general result, and the recognition of that result, are as follows:

Everywhere in space I can think of three principal directions intersecting one another at one point at right angles; or,

The general surrounding space can be determined and measured by three principal directions at right angles to one another.

The form of each object which fills space can be defined according to length, width, and height or thickness.

The thoughtful observer of language is here met by a connection of the designation by speech and the mute perception of fact; wide and widening—thick and massive *—length and lengthening.

The form of a body, the actual form of an object, is determined by the form, position, number, size, union or separation of the surfaces, edges, or corners. Hence the indispensable requirement for the abstract (drawn away), reflective, comparing consideration of all the just-named references and relations to the filling of space; hence introduction
        into the science of space,
        into the science of form,
        into the science of number, and
        into the science of size,

---

* The German words for thick and massive (dick, dicht) are also similar.—Tr.

but also further into the science and art of $\begin{cases} \text{language} \\ \text{and sign.} \end{cases}$

The consideration of the fourteen solids introduces you now to all these above-named, particular, independent, and isolated considerations and exercises, again and again, if you will only connect the continually progressive exposition of the fourteen solids with what has been presented and stated in the Education of Man and in other essays found in this and the former volume,* and, having done so, if you will then, proceeding from the cube, analyze it in its individual parts and will elevate their individual perception, as has been already said many times, to general conceptions—if you will therefore descend from cubes to tablets and surfaces and from the edges of the cube to lines and sticks.

I entirely lack both time and space to demonstrate all these facts individually in this place, and a mere inanimate verbal perception would be of little use. I must here request you to continue to develop independently and intellectually what the kindergarten has given you on this subject.

You can carry on the instruction in numbers in their rudimentary compass from the knowledge of the individual numbers and their difference up to

---

* Vol. xxx, International Education Series.

teaching their relations and proportions; from the stage of perception up to that of intellectual conception. The instruction concerning the form and size of solids, surfaces, and lines likewise proceeds from the perception of them to the intellectual conceptions of form and size, and their inner reciprocal relations to one another, but *this* instruction also should be within the general, fundamental compass.

The perception and comprehension of form, size, number, in general of figure, lead to the perception, comprehension, and knowledge of the surrounding world—in short, to the consideration of the outer world.

The consideration of the outer world in its primary conception constitutes a principal subject of the connecting school. Here also I can refer only to the Education of Man, by Friedrich Froebel, although in the more than a quarter of a century which has elapsed since the book was written and published the mode of treating this subject has been manifoldly improved and simplified. This consideration of the outside world leads in a very remarkable way (which has not yet been completed and carried through in the education of human beings) into the linking together of the activities and vocations of man, and even into the history of human development.

The consideration of the outer world leads just

as remarkably to the perception and comprehension of the province of language as audible and (by means of writing) visible representation of the outward and inward world of man. It includes within itself the whole fundamental written and spoken language of our mother tongue in a corresponding compass.

The tone and rhythm (law of movement) of words and sentences, of the speech-whole, is here again linked with language in the elementary song, and the elementary exercises in singing.

But song leads the child back again to Nature, and thus are developed from the general consideration of the outside world the actual contemplation of Nature and natural science in their rudimentary compass, and, in particular, as an important germinating point and starting point—the science of plants.

With the science of plants is organically and vividly connected the science of the surface of the earth.

For many plants are fond of the water and bestrew the shore of the brook and of the river, and encircle the sources of both. Many plants would fain adorn the meadows and vales. Many love the clear, airy, and fragrant height of the hill and mountain. Many like the vicinity of man, and many, the simple, hidden, woody vale. The vessel that crosses the ocean brings us many from distant parts of the world. The steamboat on the river, the canal, the railroad, etc., bring many. They are in the home, the garden, the house, and even the room of each person. Thus the plants really show the way and lead to the science of the surface of the earth—the description of the earth—geography.

But our kindergarten exercises, plays, and employments are also a help here. Our molding with plastic loam and moist sand teaches us hills and valleys, and blue clay and sand represent to us brooks, rivers, streams, seas. Yes, our stick-laying makes the outlines of these by indicating the main and side directions of the brooks and rivers. But also our pricking shows itself to be of practical importance here, since it gives us six equal maps which we can utilize first as a map of rivers, then a map of mountains, again a map of cities, then a map of States, then of districts and provinces, lastly as the summing up of the whole.

With the consideration of the outer world, especially with the contemplation of the plant world, is also connected the training of the sense of color and form, the province of drawing and painting. The observation of plants, of vegetables, and especially of trees, is important for the connecting school child for several reasons. First, it shows almost the whole fundamental intermediate instruction and most completely links itself with this instruction. Therefore it leads all back to the beginning, the starting point. It also leads the child (and consequently man) to himself, to the development and use of the totality of his capacities and powers, to the recognition and fostering of his nature, to the connection and union of the great whole of the universe and of life, and to the source of life, to the oneness of life, to God who is by and in himself good, as well as to the history of the inner and outer development of humanity. This history begins in a remarkable way with the relation of man to the tree and its fruit, according to our Holy Scriptures (which even the critic acknowledges). However, in the second great principal epoch of the development of humanity, and especially in the early part of this epoch, plants and trees have again a great and remarkable importance, and play the rôle of being present with man on almost all sides of his life, uniting, teaching, admonishing, requiring, so, above all, uniting him with God: "Consider the lilies of the field."

They lead man back to and into himself, to the development, the invigorating, and the right use of his powers: "Every tree which bringeth not forth good fruit is hewn down and cast into the fire."

Concerning the relation of the degree of moral cultivation it is said: "Do men gather grapes from thorns, or figs from thistles?"

THE CONNECTING SCHOOL. 301

Concerning the action of Jesus as the mediating teacher and educator of humanity between man and the demands of eternal God, as the revealer of the eternal living truth uniting God and humanity, it is said, " A sower went forth to sow."

Concerning the relation of man to Jesus, and his relation to humanity it is said: " I am the vine; ye are the branches."

Concerning the fostering and the efficiency of the truth and its results the parable of the mustard seed and its treelike development was spoken.

But not merely in reference to religion and to the Christian religion are the tree and the whole plant world important to the child and to man. They are also pre-eminently important for us as Germans, and for our children as German children; for is not the oak, the German oak, the symbol of German national life, and of the life of each individual German? " Stand fast in the storm of life like a German oak! It stands firm," etc.

Have not all our hundreds of kinds of apples and pears (as the pomologists teach us and prove to us) proceeded from the simple apple and pear tree of the wood, by means of the cultivating care of man, connected with his observation of Nature, with observation of the original, peculiar nature of the fruit trees? What are, therefore, these cultivated fruit trees?

Answer: " Works of God, of Nature, and of

humanity." And so the individual human being also in his completed education can only become and be "*a work of God, Nature, and humanity.*"

Thus he can become what he is to be only in the undisturbed, unclouded, true *union with God, Nature, and humanity.*

With this awakened anticipation, the boy and girl are ripe for passing from the connecting school, or, if another term be preferred, the elementary school, into the school of teaching and thought, from which they will later enter the last stage— the school of vocation and life.

Therefore, my dear Emma, that you may completely fulfill what the guidance of your connecting school requires, you must always have before your eyes the image and life of a tree in its all-sided functions and relations; in that case you will certainly leave none of your duties unfulfilled, no talent, no power, no natural capacity of your pupil undeveloped.

Whether you can make the application of all that has been here said to each and every requirement of your connecting school is, of course, a question which I can hardly answer affirmatively. For I can not and may not actually presuppose that you have understood me throughout; but, as I have said, it is difficult at this stage to attain complete and all-sided clearness with all-sided application without explanatory dialogue and without connecting perception. However, you see at least by the certainty with which I express myself about what is to be required and done, that I esteem it possible to attain to it, and that I am deeply convinced of the possibility—and it is something to know that one man has overcome wavering and uncertainty, and has arrived at clearness, surety, and confidence upon the subject. It is to be hoped that this letter will, or at least can, give you this certainty.

Now for a few more isolated remarks about the box with the fourteen solids.

1. The two middle rows of compartments contain the bodies.

2. The two outer rows of compartments contain the completing forms.

3. The left inner row of compartments contains the connecting forms proceeding from the cube.

4. The right middle row of compartments contains the principal forms from the cube to the sphere.

5. The left outer row of compartments contains those forms which complete the principal forms in the left middle row.

6. The right outer row of compartments contains those forms which complete the principal forms in the right middle row.

7. The forms which complete the forms of bod-

ies can be utilized for quite new, beautiful groupings, each of which again presents much that is instructive, as well as agreeable playful entertainment.

8. The knowledge of the forms of bodies (or solids) offers rich material for interesting plays that call out the child's power of recognizing them by the senses, especially by the sense of touch when the eyes are closed or when the bodies are held behind him.

The box being turned with its opening toward you, the solids succeed one another downward from you or forward from the box in the following order:

13. The double-pointed dodecahedron.
                        The double-pointed hexagonal prism. 14.
11. The eight and two surfaced form.
                        The eight and four surfaced form. 12.
9. The four and six surfaced solid.
                        The tetrahedron. 10.
7. The six and twelve surfaced form.
                        The dodecahedron. 8.
5. The six and eight surfaced form.
                        The octahedron. 6.
3. The geometric cube.
                        The pierced cube. 4.
1. The sphere.
                        The cylinder. 2.

I hope that this representation (with the intertwining and the use of sticks indicated) will make you familiar with the complete and profitable use of this box.

I will remark in addition only that these fourteen solids, with the other forms indicated therein, lead you into the whole province of the forms of Nature and of bodies, and, indeed, into three principal divisions and series of development of these forms:

1. The production of the three surface directions at right angles to one another in the forms 1 to 10.

2. The production of an edge direction in each of the forms 11 and 12.

3. The representation of a corner direction in each of the forms 13 and 14, with which the whole production of bodies closes; but the development goes on through the forms of plants and animals, as well as through the forms of thought.

May you be able to write to me in your next letter that the trouble I have taken and my sacrifice of time have given you pleasure.

With hearty greetings from my whole household to yours,

Your faithful, fatherly friend,

FR. FR.

XI.

A COMPLETE EPISTOLARY STATEMENT OF THE MEANS OF EMPLOYMENT OF THE KINDERGARTEN.

MAN, the creature of God, appears upon earth as a sentient, spiritual being, connected, as a child, with the family by his parents, and, at first, directly connected by his mother. Thus connected, he therefore appears as a member (full of life and soul) of a family likewise full of life and soul. This conception of the child in his original nature, in his natural connections and relations, is to me the foundation and the real starting point of his all-sided, prosperous development, and of that fostering and education which satisfies the demands of his nature.

Since by such fostering, education, and development the child obtains not only his right position in regard to God, Nature, and humanity, but also that position which satisfies his own nature, and consequently obtains his right position toward and in himself, so does also the family with all its members, and with each individual of them, above all the parents, and here especially the mother.

Through such fostering she (the mother) appears pre-eminently in her true nature, her real position, and in her manifold and, to the child, important connections, for first she stands as a connecting link between her child and his Creator, the Original Source of his life—God.

Next she connects the child with her husband, his earthly father.

She is the link that joins the child with the family of which he is a member.

Through the family she unites the child to the human race, with humanity, and with each individual member of humanity.

The mother also connects her child with the Mediator between humanity and God—Jesus Christ.

Finally, and lastly, and in a special sense, she is the bond of union between the child and Nature.

The mother, as a real human Christian mother, must have a clear idea of all these connective offices, as indeed must all the members of the family. She must know and acknowledge all of them. She must manage them all in a manner corresponding to their requirements with the greatest possible insight, circumspection, faithfulness to duty, and self-sacrifice. She may not neglect or subtract from any of these connective offices. For, after all, they are of equal importance, since they collectively point toward the Original Source of all existence; because only in and by means of this all-sided connection does the child develop on all sides, and toward the greatest possible earthly perfection. The child is also to develop himself at a future time to be as much as possible a complete human being. He is even to be himself a connecting link on all these sides, and this is to be observed in his first mental activity. Thus the child is later to show himself especially as a mediator, reconciler, and redeemer between Nature and God; for, as a human being, he is connected with Nature in her many characteristics by his body, and united with God by his spirit; he is also linked with humanity by his spirit and by his body, which is animated by his spirit.

As the child can in general only be satisfactorily educated toward his destiny and the fulfillment of his vocation in and by means of the manifold connection above indicated, so he develops himself first of all—at first by the aid of the mother—in and by means of the connection with Nature.

But where mediation takes place there is always identity in some respects at the foundation of what is mediated, but the identity appears in the opposite way; or, in other words, mediation presupposes opposition in appearance, but identity in nature—that is, mediation can only take place between and with opposites which are yet identical.

If therefore the mother is to educate the child

by means of and in union with Nature, such an education presupposes in the child and in Nature that which is opposite, yet like, therefore identity and antithesis—in other words, likeness in one respect and oppositeness in another.

Likeness conditions union; oppositeness conditions a contrast.

But, according to outward condition, the child, as a human being and a mortal, appears at first as a member and part of Nature. Therefore the mother must take care that the child develops in union with Nature and in contrast with it. For while necessity rules on the part of Nature man as an intellectual being should sensibly and reasonably reflect upon what he does, and do it with deliberation, with consciousness and intelligence, with determination of mind and will. The contrast, therefore, here does not consist in the opposing or striving against Nature or in that which is opposed to Nature. But the contrast is that what was done and is done by Nature is to be done by man with intelligence, therefore with reference to the undisturbed harmonious development of his individual self, therefore with ever-constant and ever-demonstrable reference to the unclouded and unobstructed representation of his personality, therefore for all-sided physical and spiritual health.

Consequently the care for the entire health of the child is the first thing which is imposed on the attention of the mother and of all those who recognize it as a duty to take part in his education.

But, as above intimated, the child is healthy when he can express and occupy himself in all the demands of his spirit and body, in all his demands as a human being in a manner corresponding to his being and nature. Therefore, first of all, the child is to be healthy as far as it is possible.

In order that the child may attain to all-sided health, and consequently to connection between himself (his essence and being) and Nature, his body with its parts of formation, with its organs of maintenance, vivification, and will, is given to him. But with the body are given the limbs and senses, both of which, in reference to the mind, soul, and will on one side, and to Nature, the outside world, on the other, are again in contrast. The senses are given pre-eminently for the purpose of enabling their possessor to so make internal and to appropriate the essence of things as it discloses itself in outward appearance, form, movement, tone, etc. The limbs, on the contrary, are given in order to manifest outwardly the will, the desire of the mind, of the soul. Here again the arms, with hands and fingers, and the legs, with feet and toes, are in contrast to one another: the arms, etc., as tools to bring surrounding things near to one's self; the legs, etc., on the contrary, as implements to bring one's self, one's body, near to the things; therefore in the former case to move the things to one's self, and in the latter to move one's self to the things.

The hands, with the fingers and with the sense of touch to be found in their tips, are again a connection between the senses and limbs; as, on the contrary, the two arms, with their hands and fingers, and the two legs, with their feet and toes, are again in contrast with one another.

The senses, on the other hand, may be considered to correspond to the three states of coherence—solid, liquid, and gaseous—or to the mechanical, chemical, and dynamic.

The physical and dietetical treatments of the child in their whole compass depend on the body with all its parts and organs.

The limbs and senses likewise require their peculiar attention, treatment, fostering, strengthening, development, exercise, and cultivation. All this is important even for the stages of infancy and childhood.

In reference to the child's development of mind, of will, of habitude, and consequently in general reference to his moral, to his actual human development, it is necessary to observe his relation to his mother and to all those who partially take a mother's place to him; in short, to observe his behavior to all which has an arousing, beneficial, and determining effect upon him.

The power and use of the child's body, limbs,

and senses develop correspondingly with this attention and fostering, and consequently the child's activity, his impulse to activity and even to employment further develop with the growth and exercise of the powers before named. If his own little hands and fingers, the hands and fingers of his nurse, and the few objects (which are mostly but little movable) which he can grasp with his little hands, suffice at the beginning to satisfy his impulse to activity, yet very soon after his little arms and hands are somewhat developed he requires for handling an object which he can move and use quite freely and easily.

The retroactive effect on the child which becomes perceptible through his use of objects is in the beginning merely a twofold one, the testing, as it were, of the things around him as to their independent existence and also as to their free movability, and second, the exercise and the feeling of his own power. The first plaything which is now to be given to the child must also be constituted in conformity with this effect. It must be, as it were, the representative (complete within itself) of all objects existing in space, and consequently must itself contain the collective general properties of these objects. Yet further, in order to suffice for all that is required of such a plaything as the first, it must neither be able to do harm to the child, nor may the child be able to injure himself or anything else with it. But the object can neither be permitted to excite and nourish the sensuality of the child nor to awaken any other wrong tendencies of mind or heart, etc.

All these requirements are met, as has been already demonstrated and declared in many places, by a moral talisman, so to speak—the ball.

But this is not the place to further demonstrate the developing, educating, cultivating effect of this first plaything and its capacity for exhaustively satisfying this stage of the child's development. Suffice it to say that in and by the ball are represented all the essential properties, phenomena, and relations of the child's surroundings—namely, material, form and figure, size, movability and rest, all kinds of movement, all kinds of relations of space and time, and also the phenomena of light and even color. It (the ball) is the true means of exercising, increasing, and recognizing the child's own strength and dexterity. It is a means of introduction to knowledge of the general properties of objects, their use and relation; and of training the little one in language, which is important to his connection with the surrounding world and to his intellectual development. The further harmonious and intellectual development by the play with the ball, connected with tone, rhythm, and song, can actually be recognized only by one's own judicious employment of the ball in play with the child, and can only be perceived by means of one's own contemplation of the play.

It may be added that the play with the ball as a type—that is, a means of representation of the child's own inner world and as a means of apprehension and recognition of the outside world around him—attracts the young human being through his whole later youth.

Let us here go back to the earliest life of childhood.

As the child's power and the use of that power in the indicated way increases, he seeks—in order also to *hear* the expression of his activity—for a body which also sounds, or at least makes a noise, but which has otherwise the same many-sided properties as the ball. This solid is the hard, firm sphere. At the same time with the sphere, the child wishes for, and the nature of the sphere requires, an object which is its pure opposite. Without going through with the proofs of the fact on all sides and individually, I will state that this object is the many-surfaced, many-edged, many-cornered, firmly resting, firmly standing, not easily movable cube, idle, as it were, only able to be shoved and thrown, but incapable of actual rolling.

But sphere and cube are purely and manifoldly opposite to, yet like one another—a fact which may be easily perceived and yet more easily demonstrated. But Nature and the child's all-sided nature require for such objects that the connection be likewise easily perceptible and just as easily demonstrable. This connection of opposites is just what gives the constancy, and by means of this constancy the developing, educating, and cultivating effect of the child's requirements, plays, and occupations.

The connection between the rolling, round-surfaced sphere which can be easily moved and turned on all sides, and the straight-surfaced cube which is, so to speak, idle, is the cylinder, easily rolling like the sphere and standing firmly like the cube, therefore uniting round and straight.

But since the cylinder excludes the perception of the corner, and the definite turning round itself on one point, it requires and conditions again the solid which connects the three—that is, has the properties of all three: corners (points), edges (lines), sides (surfaces)—and here again uniting straight and curved surfaces. This is the cone.

With these four solids consequently the second play-whole of the child is closed. They form a whole, complete within itself, for all four have three equal lines of direction inclined toward one another at right angles. But they are different or opposite according to the manner in which these three principal directions are abiding and visible. Thus in the sphere they are changing and invisible. In the cube, on the contrary, each of them is once invisible; again each is four times outwardly visible; but likewise, also, each is four times invisibly visible—that is, it is invisible in itself, especially to outward and superficial observation, but can be made visible by bringing it to attention. From this property not only do the rest of the opposite properties of these four bodies necessarily proceed, but those properties are also in the highest degree important for the child's intellectual development, particularly for the development of his power of imagination, and for the development of the inner perception and conception of the invisible. The invisibly visible lines are here again for the child, in a particularly instructive way, the connection between the never visible lines always hidden in the interior and those which are always outwardly visible as edges.

Several other points belonging to this subject, proceeding from the comparison of these four solids with one another, and highly important for the fundamental and wise cultivation and education of the child, can not here be brought forward and carried out, but must be left for verbal demonstration accompanied by actual perception of the objects.

Still more important, on account of direct application and performance at this stage of childhood, is the intimation of the use of, and the method of playing with these solids. This method for the first and earlier stage of childhood is merely the moving of these objects, the results and phenomena of which appear so very manifold, often so unexpected and astonishing, so wonderful and almost magical, that they, just because they can not be at once, or at least not easily, explained to the child, give him such great pleasure, yet teach him that which is important in fitting him for life—to make a distinction between the thing or being and its appearance, and thus to protect himself from delusion.

By means of this manifestation of form and movement these solids and the play with them give many opportunities for the observation and consideration of form, size, and number (particularly for a somewhat advanced stage of childhood), and in many ways introduce the child into the phenomena of Nature and life around him. They are therefore, as it were, the middle point and source of the later training for school and life, as well as for the union of these.

As a sheet of printed matter, besides a lithographed plate * and a pamphlet of one hundred ball songs give directions for the use of the ball " as the first plaything and always the dearest playmate of the child," so also a sheet of printed matter, besides a lithographed plate,† gives some

---

* Printed in vol. xxx, International Education Series.
† Also in vol. xxx, International Education Series.

guidance to the use of the sphere and cube, "the children's delight." The most essential points, however, must be left for verbal communication.

Much as this quartet of playthings has been opposed and ridiculed from ignorance, unskillfulness, and want of comprehension of its nature, yet I can not refrain from distinctly declaring that I consider this gift to be as suitable as it is entertaining and instructively educating, not only for this early stage of childhood, but progressively up to the school age, and I am willing and ready to give manifold proof for the assertion, by showing results as well as by words. This assertion is especially confirmed by the personal experience of every one.

I could not omit this explanation here because it might otherwise have seemed as if I wished to ignore these criticisms on account of my inability to refute them, which, however, is done without any polemical words by the facts themselves at each simple presentation.

Let us now return to the development of the play-gifts and the modes of playing.

All the playthings hitherto considered are undivided. Yet, as before mentioned, the child likes the change to the opposite. As he in the beginning likes to use everything as a ball, so he later likes to divide everything as far as his strength allows. He also likes to build together again what is thus divided; or, as is beautifully and significantly said, "the child likes to provide himself with something to create." This impulse of the child to form and create is above all to be most carefully and constantly fostered, for the more he himself creates from and by himself with his own power and spirit in independent activity and judgment, the more will he at some future time understand himself, surrounding Nature, the Creator of both, and life in its growth and in its tranquil appearance, and the better will he understand the instruction and teaching of all this and its application to his own action.

The next plaything must therefore be divisible at least once, but on all sides and in all directions. These requirements are met by the cube, once divided but on all sides, and so into eight equal part-cubes—the third play-gift, "the children's joy."

If I am not mistaken, I have already intimated that each following plaything is necessarily presupposed in and required by the preceding; thus, the three times four edges of the cube are shown in the three principal directions of the sphere, and the six surfaces of the cube in the six terminal points of those three directions. Likewise the undivided cube, by its three surface directions at right angles to one another, shows the three planes of division, each of which goes through the middle of the cube parallel with two of its sides, which in the third play-gift appear as actual surfaces of division, and thus separate the cube into the eight part-cubes already named.

The different plays with this gift are conditioned by the different ways of separating and re-grouping it. It is therefore essential to consider two points as belonging to the nature of the mode of playing: first, it is necessary that all the eight cubes must always be used for each representation; second, that as much as possible the following form must be so developed and shaped from the preceding, that what is already formed must serve in a certain respect as the foundation and means of representation of the following form.

But the forms thus made show a threefold character, each part of which is essentially different from the others. On account of this character the forms are distinguished as forms of knowledge, recognition [apperception], or learning, as forms of life or building, and as forms of beauty or picture forms. It is essential to the development and training of the child that this distinction be retained, though intermediate forms again connect the separated.

Each of these representations is connected with the explaining word, so that the child's conceptions may be definite. Wherever it is possible the representations should be also connected with the rhythmical word and with melody, so that the child may definitely and pleasurably retain the representation; for the heart, as well as the body and intellect, is to be taken into account and nourished by these plays.

Sufficient guidance is given for the use of this play-gift by the essays and by eleven plates in the former volume,* and by one hundred little rhymes and verses for the forms of life (or building forms), seventy-one songs and rhymes for the forms of beauty (or picture forms), and twenty-two for the forms of knowledge (or learning forms), and also by a third play-gift managed by the assistance of a leaflet giving directions for its use—to which all readers are here referred.

In all these representations the three different principal directions come forth in space as length, breadth, and thickness, or height, length, and breadth, but capable of change and alteration. According to the law of development, that the later may be contained in the earlier, or that the following may come from the preceding, and consequently also according to the progressing development of the child itself, this fact determines the next plaything.

But how does this next plaything result simply and necessarily from the third play-gift? Each of the four equal columnar parts, into which one can

---

* Vol. xxx, International Education Series.

imagine the principal cube separated, appears to be divided in the third play-gift into two equal cubes by a plane which passes through the middle of the column parallel to its two end-surfaces. Each of these four columnar parts is divided, in this following play-gift, by a plane which passes through the middle of the column, but parallel to two of its side surfaces, into two equal but brick-shaped parts, the three different principal directions of which now show, also, as length, breadth, and thickness, three abidingly different dimensions which bear the relations to one another of four, two, and one. Thus results the fourth play-gift—the cube divided into eight little building blocks, " the children's favorite building material."

By the slight alteration in the direction of the plane of division just indicated, this plaything obtains a double extent of surface and length, and it can inclose a hollow space the volume of which is more than twelve times that of the eight part-cubes. The representations with this gift (now, of course, likewise according to the three different aspects indicated in the former paragraphs) obtain by this alteration an almost incredible variety by no means to be exhausted by experiment, but yet legitimate and demonstrable.

As is the case with the third play-gift, the forms represented by the fourth are of three different kinds—forms of knowledge (recognition) or learn-

KINDERGARTEN OCCUPATIONS.

ing [apperception], forms of beauty or picture forms, and forms of life or building forms.

The tablet form in which the three different lines of direction at right angles to one another come out, made its appearance in the object represented with the third play-gift; but whereas those lines vanished again as they originated, they were abidingly retained by the building blocks of the fourth gift. The oblique line of direction also appears in the further representations with the third play-gift, but alterably; yet this line of direction is also required to be abiding, as it is essential to form and building, and is therefore also indispensably retained as a form in play. This is done in the fifth play-gift in which a double advance appears, first, from the cube of the third play-gift once divided on all sides, to the cube twice divided on all sides.

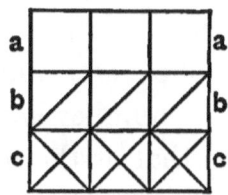

Since the law of the three now comes out in this cube, three cubes in one of the thirds of the twenty-seven part-cubes remain undivided (*a*); three cubes are each divided by a diagonal line into two triangular columns (*b*); and each of the three other cubes is divided by two diagonal lines into four columns of the same kind (c), each of which is one fourth of a part-cube, as each of the larger columns is one half of such a cube. By this division the oblique or diagonal line of direction is caused to be fixed in variously-shaped solids.

The representations with this fifth play-gift are highly important in each of the three principal kinds, forms of knowledge, beauty, and life, which have been many times named. These representations are likewise as entertaining as they are instructive and cultivating. For lack of directions and the absence of the gift itself nothing can be said about them here except that they introduce the child into school and life just as much as they strengthen and develop his mind and spirit.

As a fourth gift developed above parallel to the third, so a sixth play-gift develops parallel to the fifth. In the sixth gift, as in the fourth, the brick shape is the one determined by the law of development, and the number becomes apparent by a similar division to that of the fifth gift.

This gift has the peculiarity that by it hollow spaces and columnar erections particularly can be represented. It has the likeness to the fifth gift in that forms of beauty, the fundamental perception or position of which is the square or the triangle, can be made with it.

Being necessarily and manifoldly conditioned, the seventh gift results from the fifth, since the cube is divided three times on each side, so into four times four times four, or into sixty-four cubes. Several of these part-cubes are again divided, from the middle and through the middle, into oblique-surfaced equal parts, one half, one third, one fourth, one sixth. By arranging them together in reference to a common center the most important polyhedrons, the octahedron and dodecahedron, can be represented as if their germs existed in the interior of the cube, and, as it were, developed from it. This play-gift is highly important, since some polyhedrons, although at first only in outward form, appear as if conditioned in and required by the interior—the middle of the gift.

The seventh play-gift goes side by side with the eighth, which is related to the seventh as the sixth is to the fifth and the fourth to the third. The further development of the eighth gift is not given here.

As a review of the whole, and for the harmonizing and understanding of the annexed review, there need only be said that, by the industrious use of the body, limbs, and senses, and the plays which proceed from it, the first series of the employments and plays of children is given.

The ball, which can be used in so many ways, forms the first series of children's playthings.

The sphere, cube, cylinder, and cone form together the second series.

The third, fifth, and seventh gifts form the series of cubical forms and the forms used in play which evolve from the cubical. They thus form the third series of children's playthings.

The fourth, sixth, and eighth gifts form the series of the brick shape, or the fourth series of children's playthings.

With the development of the solid forms and their division, which are especially the subject of the last-named series of plays, the conception of the surface, as independent of and abstracted from the solid, now also appears, and, like the former solids, shows itself as an attractive plaything for children, especially if color be connected with it as emphasizing the independence of the surface. Thus necessarily appears a new division of the developing, educating plays for children—plays which arouse the child's creative power. These are the tablet-formed surfaces, which are again divided into four different series:

A. The series of square tablets, consisting of eight square tablets of two colors each.

B. The series of right-angled isosceles triangles, consisting of five gifts: the first of four, the second of eight, the third of twelve, the fourth of sixteen, the fifth of sixty-four right-angled isosceles triangles.

C. The series of equilateral triangles, consisting of five gifts: the first of nine, the second of eighteen, the third of twenty-seven, the fourth of thirty-six, the fifth of fifty-four equilateral triangles.

Finally,

D. The series of right-angled scalene triangles, consisting of one gift with fifty-six such triangles. With this gift also obtuse-angled isosceles triangles can be represented, thus completing the representation of all the principal kinds of triangles.

The derivation of these plays from those before existing can not here be carried through, yet it is quite clear to the thinker and to him who has a conception of the manifold in its unity.

In this creative means of employment the laws of development and unfolding from the inner, from the opposite through the connecting forms, are especially prominent in the so-called forms of beauty. They likewise show clearly—proceeding from simple unity, and going on by opposites and their connections—the return to unity, in this way running through the necessary series of experiences, the indispensable condition of knowledge, the necessary conditions, as it were, of becoming conscious and of later consciousness.

These plays show in various ways how movement can develop the contrasts from the legitimate regulated connection. They are especially important on account of the classical, normal, sufficing spirit applicable to the life of Nature and of man which is expressed by them, and because they clearly place before the child, one might say within his grasp, the laws of their origin in their results, as well as in their beginnings; and thus conduce to the regular, harmonious, intellectual, and uniform development of the little one.

With this gift, as with the four preceding ones, in the representations the threefold order of forms of life, beauty, and knowledge is rendered prominent, and here the exercises in color, especially with A, are added. The relations of form, size, and color are connected as in two of the preceding gifts.

With the development of the tablet and surface form and their further division, there now appear especially the conception and perception of lines. The single sticks being, as it were, embodied lines, show themselves as an attractive plaything for children, and form a whole new division of the developing means of play and employment.

The plays with the straight, unconnected sticks show again the greatest variety.

First, they can be form-plays—that is, representations of figures and objects connected with number, in which form preponderates and number is subordinate. What can I not represent with three sticks of equal length!

Second, they can also be genuine number-plays, connected indeed with form, but where the number is the principal consideration. For instance, in how many different positions and sequences can the series of number from one to three be laid with sticks of equal length? Behold:

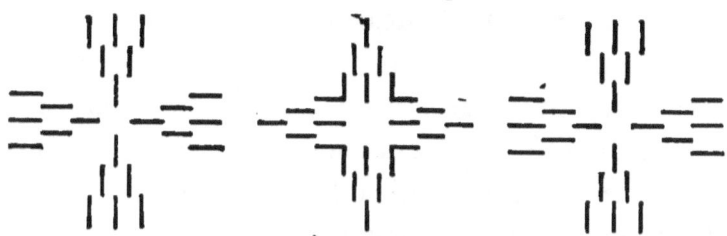

Third, they can be plays for the perception and comprehension of linear and surface sizes, especially for the relative position and inclination of the lines; for example, connection of vertical and horizontal lines (sticks) where the horizontal is just as long, or two or three times as long, as the vertical; for example, , etc. Or so that the horizontal stick is one half, one third, etc., of the vertical. For example:

In like manner with squares, or rather rectangles:

Also the comprehension of oblique lines—that is, sticks—which may be compared to trees bending lower and lower.

The actual childlike way of playing can not of course be given here, where we merely treat of the connected presentation of the means of play. With this play is included special training of the eye, and of the power of measuring by the eye. Here also, particularly with the form-plays, appear again the laws of formation and development from within of different forms, conditioned by the interior, which are recognizable by the child's power of perception and comprehension, and on that account are very suggestive and important; for experience and the repeated expressions of the child teach that he seeks to recognize in every phenomenon its inward cause.

As a connection of the opposites, there develop, fourth, representation plays and forming plays with sticks of a certain length and position; for example, with vertical sticks one square long:

Since now through all this the child's power of representation and observation, later of speech and hearing, is developed, so also through the tendency to imitation, through the necessity and law of this tendency, is developed in the child the impulse to connect what is audible and what he has heard especially words, with visible signs written and printed. Thus proceed from the first and fourth ways of playing, fifth, the letter and word laying, the writing and reading plays; at the foundation of which are the mostly straight-lined capital Roman or Latin letters (I, N, V, M, etc.). In order to lay the round forms, such as P, straight sticks are nicked.*

The plays with the unconnected sticks were, according to this statement, imitation plays, or plays of formation, number plays, plays of size and relation, purely form plays, and finally letter and word plays, or, as it were, writing and reading plays.

As now the color exercises are particularly connected with the tablets, so the sign plays and the actual writing and reading exercises are connected quite simply with the sticks.

The points of connection at which to render prominent, to observe, and to practice the singing tone are so numerous that no special prominence has been given to them before. This also has been the case with the points of connection for the observation of Nature and life, for the exercises in speech and language, and for the applications to the development of the feelings of order, right, morality, even of religion, to the training of will and charac-

---

* The rings were afterward used.—Tr.

ter, etc. The development, strengthening, and training of the power of action, of the family feeling, and of the feeling of community or rather the general feeling, appeared manifoldly and often, so that no special prominence has been given to them in the preceding pages. However, the leader of children must have a clear and definite idea of all these references so that he [or she] can render them prominent and arouse and foster them in the child whenever opportunity offers, and the capacity for receiving them shows itself in the child. This remark is one which applies to all these instrumentalities of childish play and employment.

In the same way as the perception of lines came out through the development and division of the tablet and surface forms, so also appears, in and with the perception of lines, the conception and perception of points, partly as a perception of fact, partly as mental perception, and as material and means of play for the children. The points form again a new division consisting of these plays and means of play. Objects compact by themselves and having the characteristics of the point—seeds, pebbles, little leaves, even bits of paper, etc.—can be here employed as representatives of the point. This is therefore the point of connection for the collection of natural objects, fruits, etc., as further means of play as well as for the seeking out and separating of objects which are of a like kind from those which are unlike and manifold.

Consequently we have come to the complete division and separation, to the complete dissolution, and, in certain respects, to the spiritualizing [rendering less material] of the solid, since, by the analysis of the solid into the surface, line, and point, we have followed, as it were, the development of a tree, of a plant from the germ—the seed—through the formation of stem, branch, and twig, through the development of the leaves and blossoms up to that of the anthers and stamens, and even to the development and scattering of the pollen. Hence we must now in the opposite, yet like manner, return on our path of collecting, combining, and uniting, till we again reach the first unity.

It is highly remarkable that the dividing, the scattering, immediately conditions unifying and collecting; that the separating requires at once a cohering. This coherence (this placing of the parts together to form a whole) can now, as was indicated in the former division, be performed in various ways, as, for instance, by means of a pin which is stuck into a suitably soft surface (such as can be pricked), a firmly stretched cloth or cushion, :·:

The inspection of round bodies, such as beads of different colors and sizes, also belongs here.

The connection of the points to lines, and through them to shaping of forms, is particularly expressive, collecting, and unifying. This is best done by a little ingenious device, by connecting with one another and combining the perforations in the so-called perforated sheet. This perforated sheet has the astonishing and new peculiarity that each product may be manifolded (from six to twelve fold), as the sheet when folded together many times is inclosed in another, on one side of which is indicated that combination of points by lines which is to be represented on each of the leaves.

Three such series of pricking sheets are now prepared. The first series is for the smaller, the second for the middle, and the third for the larger children. The latter series is prepared in accordance with the strict progress from the easier to the more difficult, and from the simple to the complex, which makes it a preparation and previous training for drawing and for instruction in drawing by the cultivation of the measuring power of the eye and that of the sense of form thus effected.

With the pricking is also connected the representation of letters—as initials of names—and of words in little sentences, therefore writing and reading. After all this is manifolded by the pricking, that which is represented by it may be again brought into prominence by colors; thus practice in colors is connected with the pricking. It is important that the things produced should be used as little presents, so that the child may not merely receive.

As in the foregoing the points were again united into lines and thus to figures, so in like manner single sticks, when somewhat extended in breadth, may be combined as strips, at first to surfaces, and later also to solids. The first can be done in a fourfold way:

a. By interlacing somewhat wide and flexible wooden slats which are alike in width and length. The interlacing is either voluntarily taken apart so that different forms may be made by the same sticks, or the ends of the sticks may be fastened together by some means—glue, paste, or wafers. In this way model forms may be made for other children and institutions.

b. By so-called intertwining of strips of paper folded twice, or even stiff ribbon and cord. The results of the intertwining are similar to those of the interlacing, but more lasting and firmer; for instance, in the following figure:

    c. By interlacing, or rather weaving, of paper strips. It is best that these should be of different reciprocally completing colors—that is, complementary colors. In this way the different surface forms may be again represented. The result of this activity may be employed, like figured paper, to make or beautify the most different objects— portfolios, writing books, needlebooks, cuffs, collars, napkin rings, etc., box covers, covers for glasses, cushions, etc.; thus again for various little gifts and presents.

    The use which is made of these little products, as was mentioned above in regard to the pricking, is very important to the civilizing, to the nourishing of the child's being and mind; for I consider the fact that many children receive so much and can give hardly anything from their own little productions, to be one of the most essential causes of the frequent retrogression of childlike love and sensibility. This statement must here suffice as an intimation of one of the most important points of education.

With increased development of power and under precisely stated relations, the weaving of mats with straw, etc., and even basket making, are connected with the weaving. With the above-mentioned pattern leaves, etc., this may become at once an industry for poorer children.

d. The fourth combination is that of slender-pointed sticks connected by softened peas:

1. To form simple figures, triangles, and quadrangles, but especially the triangles, in accordance with fixed laws. These simple figures can then again be grouped together in the most manifold ways to form structures, but likewise according to fixed laws. This grouping together is again a most excellent cultivation and preparatory training of the eye for drawing.

2. To make equal-surfaced and otherwise symmetrical solids which can be seen through, being, as it were, only represented by the edges in outline; and—

3. Also to construct buildings, house, room, and household furnishings, which can also be seen through.

By the inner combination and composition of equal-surfaced and symmetrical bodies the development of each following form from the preceding, and finally from the original, fundamental form, may be represented in such a manner as to be capable of being seen through and capable of being seen into.

This occupation is extremely important for the insight into the innermost nature of the development of forms and solids gained by it, as well as for the impulse to creative activity. It is also particularly important for the development and cultivation of the mind by the anticipation and investigation of the unity of all that is formed from and through life. Hence it comes that the children dearly love these occupations.

Three remarks may be here permitted:

First, the cutting from wood of all kinds of little furnishings for house and household, such as ladders, sleds, sawhorses, etc., is connected with the splitting of the sticks from flat bits of wood and tablets, and with the sharpening of the sticks.

Second, by these employments, models, solids, and objects can be made for other children and institutions. Thus results a means of industry and support and, at the same time, of education and training for poorer children. This fact has already been mentioned several times.

Third, the children are hereby enabled to make little presents to those whom they love.

The combination of surfaces forms a new but natural sequence to these employments. This combination takes place first of all with paper surfaces, which, by creasing and folding, form solids, or rather the outlines or superficies of solids—here boxes of the most different kinds, with and without covers, there equal-surfaced, symmetrical bodies, and buildings. This creasing and folding is used particularly with a later occupation, that of cutting. By combining surfaces into the form of symmetrical, rectangular solids, all cubical sizes can be represented; for example:

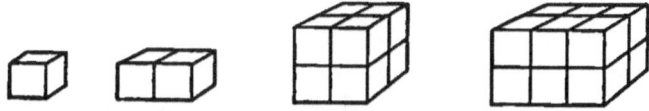

Of course the way of fastening the sides rises likewise from the easier and simpler to the more difficult and complex. At first the fastening is done by simply bending the material, then by interlacing and closing together, finally by gluing or pasting.

As this occupation advances according to rule, it also is entertaining, enjoyable, cultivating, and instructive for the child.

With this occupation are connected the actual paper work and the work with pasteboard for older and more cultivated children, particularly boys; and the remarks which were made above in respect to children's gifts and to the cultivating effect of these means of industry for poorer children are therefore also of value here.

A further new and large division of the employments of children, as entertaining as it is instructive and useful, is the change of form:

A. Without diminishing the quantity of the material, and here—

a. At first with flexible thread lines—the twining into cords and other objects.

b. With proportionately connected stick lines, as, for example, with the usual inch sticks. Here especially representations of the different angles and figures by changes.

c. With flexible surfaces, such as paper, by creasing and folding, different forms and objects are made with one and the same square surface, or, what is the same, with several square surfaces of equal size.

d. With soft flexible material—wax, loam, or clay. Modeling is included in this occupation.

B. With the diminishing of the quantity of the material:

a. Cutting of different forms of beauty from square surfaces of like size according to fixed, precise law, generally the law of bringing out the visible and that which can be made visible from the invisible or, what is the same, from the hidden visible. The forms are here:

1. With straight lines.
2. With curved lines.
3. With both straight and curved lines.

b. The cutting of different objects:

1. Free-hand cutting of the objects of the house and household, natural objects, and the objects of social life.
2. The cutting out of already depicted objects —animals, furniture, people, dresses, soldiers, the objects of trade, landscapes.

Here takes place the unfolding of the whole human department of social life, and of the life of Nature—morals, good behavior, sensibility, the right, the view of and insight into life, history; also, on the part of the child, the use of all the unfolded powers and qualities of his mind as well as his heart, of his speech as well as the activity of his body, limbs, and senses.

With this employment and play the busying of the children for this stage of childlike development terminates, as it embraces and comprises the whole life of the child in every respect; man comes into consideration in all his relations to God, Nature, and humanity.

This is shown by the extended second play-gift or " the sphere and cube with the solid forms derived from them according to necessary laws."

c. The change of solids made of soft material that can be cut—for example, clay, loam, potatoes, turnips, cabbage stalks, or soft wood—and first of all the change of the cube by cutting off (1) the corners, (2) the edges. Here is specially revealed the endeavor of the child to perceive the inner in the outer, to represent the inner by the outer; and, in reverse order, to find unity in manifoldness and to develop manifoldness from unity. The way in which this busying of the children leads to the higher and true (religious) comprehension of Nature and of their own life is clearly evident. And consequently here is also a complete close.

This again leads us back to the consideration of the plays with the solid and firm round, and here, as a connecting point with the former, first of all:

A. To the undivided sphere and to the plays and employments with the sphere in different sizes and numbers.

B. To the divided round. First of all,

a. Spheres are divided:

1. Concentric with the surface, thus giving half spheres and spheres within others.

2. Parallel to one of the largest circles, giving disks.

3. Through the three largest circles, intersect-

ing each other at right angles, thus into eight equal four-surfaced bodies.

b. The cylinder is divided:

1. Concentric with the cylindrical surface, thus into cylinders of different sizes.

2. Parallel to the upper and lower surfaces, thus into equal-sized disks.

3. Through the two largest division planes intersecting each other at right angles, thus forming four columnar bodies, the sides of each including two plane surfaces and one curved.

4. Into circles or rings from number one.

c. The cone is divided again:

1. Concentric with the curved surface.

2. Parallel to the base into disks.

3. Through the two division planes cutting through the axis at right angles.

4. According to the conic sections.

From the connection of the round (thus originated) with the straight; and—

1. From the outward connection of the round with the straight proceed the building and laying forms, and

2. From the inward connection of both, the roller, the wheel, the barrel, the wagon, the carriage, etc.

With the modeling as an object of play is, therefore, connected

I. The province of mechanics.

II. The province of the contemplation of Nature and introduction into it.

III. The province of human and social life.

But the round also finds its educating application in play in the conception and connection of planes and lines.

From all this now proceeds the free connection of each favorite material in favorite forms in the most manifold way and for the most different objects of play, employment, entertainment, and education, as well as other objects (the fostering and observation of human life and human relations); for instance, for gifts of friendship, as has been already many times mentioned.

With these is further connected, as an educating employment for play and entertainment, first of all the collecting of natural productions (pebbles, leaves, flowers of the most different colors and forms) to which the child so early inclines, in order by this collecting to exercise his power of observation and comparison, as well as to extend his objective knowledge, but, above all, to procure for himself a knowledge of the objects of Nature.

The collecting of objects—the flower, the plant, the bug, the caterpillar, etc.—leads to the care over them for growth and unfolding, therefore to the fostering of life. The delightful impulse to such fostering likewise shows itself very early in the child as soon as he has even an anticipation and perception of it. All this has an instructive effect on the parents, and in general upon all who exert an educating influence on the child. The gardens of the children belong here essentially.

Lastly, humanity, in the shape of the child, is even for the child the most satisfactory object of play and the most enjoyable playmate, as through the child all expressions of human life become material for play, and, as it were, the bringers of play. From this fact proceed the voice and singing plays, the word and speaking plays, the movement and representation plays, especially the representations of man in the most diverse stages of development, in the most varied relations, vocations, businesses, and efficiencies. They are in general, therefore, the plays of observation, comparison, and consequently apperception; plays for the exercise of thought, for the fostering development and cultivation of the reason, the intellect, the head and heart, manners, and modesty, as well as of morality and the highest union of life—the greatest fostering and observation of life in all relations. For the child thus represents, by and by means of himself, his innermost unconscious life, as yet unknown to him. He absorbs the inner and innermost of the surrounding joint life. He mirrors this life in himself, he compares both spheres of life, finds what is common to both, what is alike and what is united in both. He thus develops, educates, and forms himself for true, all-sided, intimate union of life, and so for the understanding of life, for insight into life, and for the ruling of life (as far as this is possible at the child's age and at the different stages of development). This is the actual and highest aim of this play-whole, of this means and way of playing, of this course of play. As a higher spiritual bond passes through and pervades the whole, so the child in the plays makes the discovery that he can only arrive at insight into and practice of this united and constant whole through constant connection. In these three, and in the presentiment thereof recognized or presupposed in the child, lies the developing, educating, forming, high effect of these plays. This high effect lies in the still deeply slumbering, yet already active presentiment in the child that only the transitory and visible can by connection lead to the becoming objective, to the becoming visible, and consequently to the comprehension of the single and invisible that abides in itself, therefore to the solution and perception of that which is inward, spiritual, and directing.

In this anticipation the child's impulse to imitation, the tendency toward and the power of reproduction have their foundation as well as their aim;

and in this anticipation is revealed the nature of the plays and ways of playing here presented.

By means of all this the whole life of Nature and of man, the nature of all things, and above all that of man (as being whole and single, dissolving all opposition and contrariety, and consequently harmonized), are clearly revealed to the child in the mirror of his plays.

THE END.

www.ingramcontent.com/pod-product-compliance
Lightning Source LLC
Chambersburg PA
CBHW020309240426
43673CB00039B/750